Soft Matter: A Very Short Introduction

VERY SHORT INTRODUCTIONS are for anyone wanting a stimulating and accessible way into a new subject. They are written by experts, and have been translated into more than 45 different languages.

The series began in 1995, and now covers a wide variety of topics in every discipline. The VSI library currently contains over 650 volumes—a Very Short Introduction to everything from Psychology and Philosophy of Science to American History and Relativity—and continues to grow in every subject area.

Very Short Introductions available now:

For more information visit our website

www.oup.com/vsi/

Tom McLeish

SOFT MATTER

A Very Short Introduction

OXFORD
UNIVERSITY PRESS

Great Clarendon Street, Oxford, OX2 6DP,
United Kingdom

Oxford University Press is a department of the University of Oxford.
It furthers the University's objective of excellence in research, scholarship,
and education by publishing worldwide. Oxford is a registered trade mark of
Oxford University Press in the UK and in certain other countries

Published in the United States of America by Oxford University Press
198 Madison Avenue, New York, NY 10016, United States of America

British Library Cataloguing in Publication Data

Data available

Library of Congress Control Number: 2020935145

ISBN 978-0-19-880713-1

Printed in Great Britain by
Ashford Colour Press Ltd, Gosport, Hampshire

In memory of two wonderful creators and teachers:

Sam Edwards
Pierre-Gilles de Gennes

*and with thanks to the global community of soft matter
scientists and students who have given me so much.*

Contents

Preface

Any addict, like myself, to the 'Very Short Introduction' series entertains a dream that one day they might be asked to contribute one. So it is with particular joy that I have been able to write this introduction to soft matter, and to acknowledge the attention to detail, patience, and advice from Jenny Nugee and Latha Menon at OUP, which have been outstanding. That is only one of the many privileges that this little book represents, however. To have witnessed, and to have been involved in a small way in the gestation and formation of an entire field of science is an experience not granted to many. Nor is the engagement with a topic that has the interdisciplinary reach of soft matter, from physics to chemistry, chemical engineering to biology. It is equally uncommon for a single field to bridge science going on in university and industrial laboratories with such ease of communication and collaboration. Finally the very place that soft matter holds in the recent history of science, and the way that it leads us back to our roots in the curiosity about the world immediately around us has led to fascinating discussions with historians and philosophers of science. To the many researchers across this rich and international community who have taught me patiently over the years, I am very grateful indeed.

I am also fortunate to have received the type of wisdom that only ever seems to reside in the first generation of pioneers within a field. Particular gratitude is due to the inspirational advice of Prof. Pierre-Gilles de Gennes, Prof. Dame Julia Higgins, Prof. Walter Stockmayer, and above all Prof. Sir Sam Edwards. From the other end of the generational spectrum, those colleagues who have worked with me as post-doctoral fellows and PhD students have taught me far more than I them, for which I am extremely grateful.

The immediate context and inspiration of this book is the EPSRC Centre for Doctoral Training in Soft Matter and Functional Interfaces ('SOFI-CDT'), based at the universities of Durham, Leeds, and Edinburgh, and partner industries (https://www.dur. ac.uk/soft.matter/soficdt/about/). To the many staff and students who have grown the 'SOFI' community over the last few years I am very grateful, but particular thanks are due to the students who gave me their advice, diagrams, and editorial eyes on particular chapters: Peter Wyatt, Rebecca Fong, Vishal Makwana, and Vanessa Woodhouse.

I am grateful to my new academic home at the University of York for giving me the time and space to complete the project.

List of illustrations

Chapter 1
The science of softness

Lucretius, the Epicurean poet and philosopher of the 1st century BC, wrote a remarkable compendium of natural history (*De Rerum Natura*—On Natural Things), which introduces the power of conjecturing that all substances are composed of tiny, invisible 'atoms', or 'seeds':

> Wine, we see, will flow on the instant through a sieve, but oil is hesitant and slow, either because its particles are larger, or else more hooked and tightly intertwined, and this is why they can't be pulled apart so quickly into separate single atoms that seep through single openings, one by one.

The Epicurean tradition sought to explain the observable properties of material things as arising from the structure of their atoms. Though centuries before the experimental and theoretical methods of modern science provided incontrovertible evidence of the existence of atoms, and the larger and more complex molecules which they build, the ancient atomists had grasped a fundamental explanatory key to the world.

Lucretius, and others of the school such as Democritus, were particularly interested in the behaviour of matter as it deforms and flows. The sluggishness of honey is a visible clue, pointing to the invisible and underlying nature and interaction of its atoms.

Whether human observers reflected scientifically on it or not, the technology of pliant and deformable matter has an even longer history than the atomic hypothesis. The inks used by the Babylonians of antiquity and the rubber latex of ancient Central America are as indicative of intriguing material properties as the wine and honey of classical Rome.

The Epicurians were also correct in suggesting that flow properties of materials are pointers to their underlying structures. Heraclitus, a much older pre-Socratic philosopher, is credited with the aphorism 'everything flows' (Greek, *panta rhei*—παντα ρει) as a metaphor for the continual dynamic change inherent in matter. The Greek for flow, *rhein*, has given us the name for the science of flow and deformation, *rheology*. The different rates at which fluids flow under the same force—the sluggishness of honey and the quickness of water—was, for the ancients, and is in today's experimental laboratories, the fundamental characteristic of soft materials. Rheology has now been transformed into a highly quantifiable tool, accompanied by modern techniques such as microscopy and light scattering that allow direct visualization of the tiny structures within its materials of interest. 'Soft stuff', by definition, deforms and flows; hence rheology has also become part of the broader interdisciplinary field of *soft matter* that has emerged over the past three decades.

Soft matter science is special. It begins with the commonplace and familiar, yet it can take us to the frontiers of the exotic unsolved. Who would suspect that the slow drip of the hevea tree's latex would connect to the 20th century's theoretical physics of 'field theories'? Or that the discovery of carbon-based inks in the ancient world would lead, via the work of a botanist peering at the constant motion of particles within plant seeds, to Einstein's work on heat and to the final convincing proof that atoms exist? Who would have guessed that the ancient discovery of soap-making would lead to an understanding of the self-assembled structures of the cells in living organisms themselves? This Very Short Introduction concerns deep ideas that pick up momentum on timescales of centuries yet also constitute some of the great cultural threads of our own time. The science of soft materials brings new insights into the world we see, hear, and feel around us. The lustrous whiteness of milk, the colourful swirls on soap bubbles, the stringy stickiness of melted cheese, the extraordinary extensibility of rubber, the switchable states of 'liquid crystal' displays—these everyday wonders may be familiar, but it is quite another thing to understand them. Such immediate sense impressions have nevertheless always been the starting point for science, so it will be those familiar properties that set the narrative of this book: the phenomenon of 'soapiness' leads to the science of 'self-assembly' and the experience of 'sliminess' to the idea of giant 'polymer' molecules (Figure 1 shows some different examples of soft matter, together with microscopic images of their underlying structures).

From the 1990s onwards, terms such as 'soft matter' and 'soft matter physics' began to appear with increasing regularity in conference announcements, review articles, and books. The new language reflected a cohering of previously much more fragmented scientific communities. 'Soft matter' had begun to subsume sciences that had previously gone by the names of 'colloid physics', 'polymer physics and chemistry', 'liquid crystal science', and more. The new term indicated not simply a merging of research programmes but that the whole represented more than the sum of those parts—the recognized emergence of a new field of science that drew on shared conceptual and experimental foundations. So materials falling under the labels of 'colloids', 'polymers', 'liquid crystals', 'self-assembly', 'membranes', 'foams', 'granular materials', 'biological materials', 'glasses', and 'gels' now find a common scientific home. Conversations between these sub-fields as well as within them are in consequence now commonplace.

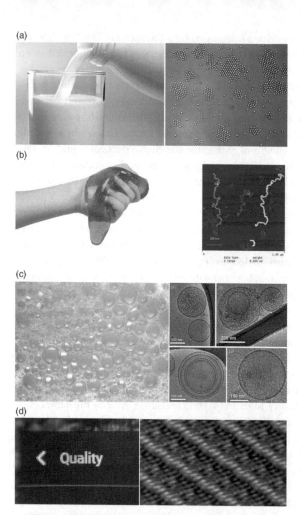

1. (a) Milkiness; (b) sliminess; (c) soapiness; (d) pearliness—the
material manifestations of the soft matter examples of colloids,
polymers, surfactants, and liquid crystals. In each case, the ordinary
scale pictures are accompanied by microscopy images of the
'mescoscopic' or middle-sized structures underlying the
soft materials: (a) colloids—particles; (b) polymers—strings;
(c) foams—membranes; (d) liquid crystals—molecular rods.

Another contribution to the emerging history of soft matter science has been made by fruitful engagement between industries and universities. Many pioneering individuals have held posts in both, and many ideas and discoveries had their origin in industrial, rather than academic, laboratories. In the case of the science of plastics, for example, Du Pont in the USA and ICI in the UK were birthplaces for the synthesis of new materials and early theories of how the giant string-like molecules of 'polymers' are formed. Fundamental phenomena, for example the relation between the viscosity of melted plastic and the mass of their long-chain molecules, have often been first observed in industrial conditions. Even use of beams of sub-atomic neutrons—the technique known as 'neutron scattering'—to elicit the first polymer structures was driven from these large industrial laboratories. Soft matter has illustrated how the fields of technology and science are far more closely entangled than the common phrase 'technology transfer' would suggest.

Furthermore, the nature of experiment and ways of theorizing in soft matter raises profound questions of where 'fundamental' scientific ideas really lie. The American physicist and Nobel Laureate Philip Anderson, in a landmark article entitled 'More is Different' published in the 1970s, described how structures can 'emerge' at length scales much greater than those of atoms and molecules, but which are just as 'fundamental'. Soft matter provides many illustrations of Anderson's claim that the notion of fundamental physics should not be tied to any one scale of length or energy, and that while 'reductionism' (the explanation of the behaviour of a system purely in terms of that of its smallest constituents) is an essential tool in science, it cannot be the whole story of how we understand the world. Nature is built from many components, but fundamental novelty arises also from the way they are assembled hierarchically. A few dozen atoms can build small molecules, which can in turn be assembled into giant 'macromolecules' or 'nanoparticles', whose properties now depend not on their tiny building blocks but on their shape and structure as a whole.

The development of soft matter science was propelled by a combination of communication within the scientific community (through conferences, research departments, and journals that attracted scientists from more than one sub-field), intrinsic conceptual overlap and commonality, and visionary leadership from a small number of pioneering scientists. Pre-eminent among those were two theoretical physicists working in the latter half of the 20th century, Pierre-Giles de Gennes in France (later a Nobel Laureate in physics), and Sir Sam Edwards in the UK. Both realized that broad conceptual frameworks and powerful theoretical techniques from other areas in physics could be applied to soft matter systems, and that as a result, simple and deep structures could be perceived beneath what had previously seemed a disparate collection of very complicated materials. The conceptual leaps were considerable: ideas from fields dominated by the structures of quantum mechanics required translation into systems dominated by thermal physics, whose tools and techniques appear at first sight to be quite distinct. The background of Edwards (a former student of American Nobel prizewinner Julian Schwinger) was quantum field theory—the physics of interaction of light with matter; while de Gennes (a former student of French physicist Jacques Friedel) had worked on superconductivity—the bizarre phenomenon of zero electrical resistance that appears in some metals at very low temperatures. The theoretical models they forged to describe the soft matter physics of polymers and liquid crystals respectively were written in the language of mathematics adapted from their earlier work, but the tangible soft matter examples opened up other, more visual and diagrammatic ways of understanding. These, rather than the formulas, will provide our conceptual ways into soft matter.

Chemistry proved as essential an ingredient to the new science of soft matter as ideas and techniques from physics, together with a renewed conversation between the two sciences. Chemists brought not only the careful synthetic construction of the remarkable giant molecules and assemblies of soft materials but also ways of thinking about and characterizing the subtle forces between them. Arising from ancient technology, the imaginative leaps of individuals, and the cross-fertilization of ideas—including the final opening of a window onto the atomic world itself, imagined for so many centuries—it is not possible to tell a convincing story of the science of soft matter without visiting some more of its history in each case study.

Before we take deeper dives into some specific examples of soft matter, telling their stories, opening the experimental windows into their structures, and the theoretical concepts that have begun to turn the familiar into the understood, we first need to think a little more about what we mean by 'soft matter' itself, anticipating a handful of characteristics that are shared by the materials we term 'soft' and the science used to investigate them.

Characteristics of soft matter

The term 'soft' is used essentially in its everyday sense—but the compliance and mechanical subtlety of soft or weak solids, gels, and fluids is also the first clue to their internal structure and dynamics, for soft materials reflect a hidden inner world of constant motion.

Thermal motion

The ancient notion of atoms, and its modern development into the variously structured molecules of gases, liquids, and solids around us, is a good starting point but not quite enough to explain or understand 'softness'. One of the most astonishing demonstrations that molecular ingredients alone are insufficient to determine material properties requires a rose and a container of the extreme coolant: liquid nitrogen (the gas that, apart from the 21 per cent of oxygen, comprises most of the rest of the air we breathe, liquifies at about −200°C). The lustrous rose's petals are indeed as 'soft' to the touch of a finger as any object we might imagine and as pliant. But now cool the flower to the temperature of the liquid nitrogen by submerging it in the nitrogen for a few seconds, and when we extract it by its stalk, while there is no visible change to its colour or shape, a sharp tap from a hammer will shatter the petals into so many shards, like a fragile piece of glass sculpture (see Figure 2).

The vital difference in the molecular structure of the flower at room temperature and at −200°C is not in the chemistry but in the *motion* of the molecules. Every one of these is in constant random motion, careering now in one direction, now in another, the chemical bonds between atoms vibrating and rotating this way and that. The higher the temperature, the faster all these motions are, on average, while instantaneous values of the random velocities vary hugely. This is the chaotic molecular motion that we call 'heat' on everyday scales.

The systematic study of this essential inner motion of matter was initiated by the early 19th-century botanist, Robert Brown, who

2. A rose, frozen in liquid nitrogen, shattering into shards on impact. Its softness is a function of its temperature, not just its molecular constituents and structure.

published a landmark pamphlet on it in 1828. He described his efforts at understanding the source of a constant 'jiggling' motion and diffusion of small particles suspended within pollen grains he had observed through his field microscope. In humble, but highly responsible, fashion, Brown explained all the many possible causes of the ceaseless motion (from magnetism to currents in the fluid) that he had systematically ruled out, yet he still found himself unable to account for its source. Although the correct explanation had to wait for an equally seminal paper by Einstein in 1905, as we shall see in Chapter 2, the thermal agitation is rightly called 'Brownian motion'.

Every part of every material is subjected to Brownian motion, but where strong bonds tie the molecules or atoms to each other, such as in metals or minerals, the random thermal motions will not be able to excite large changes in structure. Shaking a steel climbing-frame in the local playground causes only a slight vibration. But applying the same force to and fro to the climbing net made of rope on the next-door apparatus will cause it to fluctuate wildly. Likewise at the molecular scale, the thermal kicks that jostle molecules around will have a greater effect on those joined by more compliant and weaker bonds. Here lies the molecular key to understanding 'softness'. Soft materials exhibit large local rearrangements of their microscopic structural constituents under thermal agitation, while 'hard' materials suffer only small distortions under the same effects of heat.

There is a way of turning this idea of fluctuations and distortions from random thermal motion into numbers—a measure of 'softness of matter'. The great 19th-century German scientist Ludwig Boltzmann gave us the 'currency conversion' formula we need to translate a temperature T (on the absolute scale where zero is −273°C) into the typical thermal energy of motion E_T that every microscopic particle will possess at that temperature. The 'Boltzmann constant' (a very small number when written in everyday units), written k_B, provides the conversion factor: just like currency conversion from dollars into pounds, we simply convert temperature T into mean atomic energy E_T using

$$E_T = k_B T$$

The higher the temperature T, the greater the mean energy of molecular motion E, delivering an increase of k_B for every degree increased (depending on the structure of the molecule the increase may be a small multiple of this number). At ordinary temperatures this energy quantum belongs at the atomic scale, tiny in everyday human terms, but it is enough to send an oxygen molecule careering off at 300 metres (m) per second (s) at room temperature (this is therefore the average speed of the molecules in the air you are now breathing). Now we have the numbers, it becomes possible to determine if a piece of molecular matter is 'soft' or 'hard'. We simply need to ask if the energy required to move each of its atoms and molecules when the material is twisted and deformed substantially (say more than by 10 per cent) is greater, or less than E_T. This is the criterion that separates glass from the biological fabric of rose petals. But only at room temperature—Brown, Bolzmann, and others had laid the foundations of the temperature-dependent science of *thermodynamics*, which would prove indispensable to understanding soft matter.

Structure on intermediate length scales

There are other recurring characteristics to soft matter beyond the essential ingredient of large thermal fluctuations in its structure. A significant one is simply located in the typical *scale* of important microscopic structures themselves. Recall how very small atoms are. If a bed bug were enlarged to the size of an Alpine mountain, then the atoms composing its leg would be the size of the original bed bug. There is therefore plenty of room at intermediate scales on which to build structures, between the size of atoms and everyday dimensions. The significant structures that determine the properties of soft matter are still tiny on human length scales yet are typically built from very large numbers of atoms. So these basic units we need to imagine in order to understand soft matter behaviour are neither 'macroscopic' nor truly 'microscopic' (in the atomic sense) but 'mesoscopic' ('meso' from the Greek word for 'middle'). Structure between the length scale of several nanometres (1 nm = 10^{-9} m, or ten times the size of atoms) to a micron (m; a hundred times this) is almost ubiquitous in soft matter systems.

To give some examples: the giant string-like molecules of polymers are long chains consisting of many thousands, or even millions of atoms. Their flexibility means that, when dissolved in a solution, they adopt randomly contorted coils whose diameters are typically several tens of nanometres in scale, a thousand times the size of an atom. The particles and droplets suspended in the milky fluids we call 'colloids' are too tiny to see with the naked eye, yet they comprise typically millions of atoms. Liquid crystals are, as we shall see, solutions of rod-shaped molecules which locally tend to point in the same direction. For an analogy, think of combing flat the hair on a dog—at every point the hairs lie in the same direction (like the local orientation of molecules in a liquid crystal). The emergent patterns of molecular direction support 'defect structures'—points or lines at which the local direction of molecular orientation suddenly changes. The behaviour of the liquid crystalline material is dominated in many ways by the mesoscopic length scale at which the defects interact (much larger in this case as well than the scale of the molecules themselves). The second images in each case of Figure 1 show schematically the different mesoscopic structures underlying the soft matter phenomena, matched panel for panel.

The dominance of these structural, 'mesoscopic' length scales in soft matter systems implies a crucial principle for understanding them. If their key structural components are themselves composed of thousands of atoms or more, then mathematical models for polymers or liquid crystals which attempt to keep track of all those atoms will not deliver insight into why they behave as they do. We sometimes say that appropriate models for soft matter systems need to be 'coarse-grained'—if we are to understand them, the fine atomic details need to be blurred out, otherwise we do not 'see the wood for the trees'. Here we encounter once more the central message of Anderson's 'More is Different' manifesto—'fundamental' structures are not necessarily atomic or sub-atomic. In soft matter they are typically much larger. The chapters that follow will highlight in each case the mesoscopic structures and the coarse-graining of the models we use to describe them.

Slow dynamics

The mesoscopic spatial scales of soft matter's fundamental components implies another characteristic property of how they move. The dynamical behaviour of polymers, liquid crystals, and other soft matter examples can be as complex as their spatial structures. Complexity in time follows complexity in space. The component molecules are moving about in random thermal motion and colliding with each other at unimaginably short intervals (in dense fluids at room temperature, the mean thermal velocity of the order of 300 ms^{-1} dictates an average collision time of about a picosecond—1 picosecond = 10^{-12} s). Yet the much larger mesoscopic structures of soft matter systems can possess 'slow variables' whose values change on much longer timescales. A common everyday example is found in foams—whether on top of glasses of beer or formed by soap froth in the washing up. The molecules within the films may be colliding on ultrafast microscopic timescales, but the foams themselves slowly rearrange and combine their cells, gradually draining away after seconds or even minutes. For another example, think of the 'silly putty' children's toy. This remarkable material can be rolled up into a ball and bounced on the floor like rubber. Yet left on its own on a table top for a few minutes and it flows into a puddle. In other words, on short timescales the putty behaves like an elastic solid, while at longer scales it flows as a viscous liquid (we say that it is 'viscoelastic'). Somehow it possesses an internal timescale—of seconds or minutes—that dictates whether it is behaving in its 'fast mode' (bouncing) or 'slow mode' (flowing). The slow flow of molten polymers, first observed in the industrial laboratories that developed early plastics, is another example of viscoelastic material behaviour.

Universality

A sort of 'super-property' of soft matter arises from the characteristics we have already discussed—the high importance of thermal fluctuations, mesoscopic structures, coarse-grained physics, and slow dynamics. It is called 'universality', signifying that the same material properties in each class of soft matter can arise from many different underlying chemistries. So, for example, the elastic strength of a polymer rubber is dependent on the density and distribution of the cross-links between its constituent polymer chains, not on the chemistry of the chains themselves. What matters is that the polymer chains are physically connected and flexible; any local chemistry that delivers these characteristics will support a network with the same physical properties. Similarly, the 'liquid crystalline' fluids of rod-like molecules all pointing in the same direction appear whenever there are densely packed molecules of elongated shape, independently of the atoms from which they are built (there are special cases such as molecules possessing electric charges that can behave differently). This characteristic of universality is of immense practical relevance. Where it holds, solving a problem for one specific example can automatically solve it for whole families of materials. It is also of deep conceptual significance, since the understanding of a general phenomenon is usually more powerful than an insight restricted to a specific example.

Experimental commonalities

The emergence of soft matter as a coherent field of study is by no means the result of theoretical ideas alone. The parallel growth of important experimental tools was also vital. Furthermore, these laboratory techniques also applied across the gamut of soft materials, in a similar way to the theoretical tools. Microscopy is the most obvious and direct example. Deceptively simple, there is a lot more to magnifying small structures than arranging a light source and some lenses lined up in front of them. For example, special techniques that collect the light emerging from just one specific plane in the object under study have overcome the confusion that otherwise presents an observer peering through many hundreds of layers of particles.

The same principles that guide optical microscopy can be applied to any wave-like probe of matter. One consequence of quantum mechanics is that all material 'particles' propagate as waves in a similar way to light. 'Microscopies' exist that exploit this 'wave-particle duality', of which the best known uses electrons to probe and image matter. Since electrons are charged particles, they may be focused by 'lenses' of electric and magnetic fields, and since their wavelengths are so much smaller than those of light, they can image far smaller details. Delicate structures in self-assembled membranes furnish an example of the details laid bare by electron microscopy (see Figure 1(c)).

The advantage of microscopy is its ability to reveal details of local structures within soft matter, but by the same token, its disadvantage is the consequent inability to sample and average such local detail across an entire sample. Finding information on averages is important when the mesoscopic structures are random—like the coiling polymer chains or the constantly rotating and diffusing rod-molecules of liquid crystals. A family of techniques that performs just such an integrative view is known as 'scattering'. Like microscopy, anything that propagates as a wave—light, X-rays, and also particles such as electrons and neutrons all make good scattering probes, depending on the materials under study. We will have more to say about scattering in Chapter 3, but here note the general geometry of an incoming beam striking a sample, which 'scatters' radiation across an arc of angles into an array of detectors (illustrated in Figure 13 in Chapter 3). An analogy of the beam might be the bowler/pitcher in a game of cricket/baseball, the fielders representing the array of detectors. Even without a visible view of the game, the frequency with which the ball is picked up by each fielder alone contains information about the batter, for example whether they are left- or right-handed.

Scattering experiments respond naturally to the scale of mesoscopic structure in soft matter. X-ray crystallographers have used the scattering of high-energy, short wavelength X-rays to resolve the tiny atomic spacings in crystals. Longer wavelengths of X-rays, cooler neutrons, or looking at smaller angles of scattering can all extend the same experimental idea to probe longer length scales than atomic spacing. These 10–100 nanometres scales are those in which the mesoscopic membranes of cell walls, the expanded coils of giant polymer molecules, or the gentle ordering of colloidal spheres (see Figure 1) all swim into view.

The feature of slow dynamical structures within soft matter motivates a further set of techniques to identify and characterize, common to soft matter, and which focus on the mechanical properties of the material. This is, after all, where the 'softness' comes from. Furthermore, there are complex connections between the structural changes at the nanoscale and the resultant flow properties of the bulk material. The corresponding techniques fall under the banner of the 'rheology' introduced at the start of this chapter, and the instruments that apply them are known as 'rheometers'. The essential operation of a rheometer is to impose specific geometries of deformation, or 'strain', on a soft matter sample, and to measure the forces that the deformation induces on its surfaces.

A common example is depicted in Figure 3, showing the simple 'shear flow' of soft material or fluid constrained between two

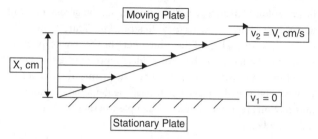

3. The arrows indicate the horizontal fluid velocities at different points in a shear cell, smoothly growing in the vertical direction from zero at the bottom, stationary, plate, to V at the top plate. The shear stress is the horizontal force per unit area on the top plate required to drive the flow.

parallel plates. The lower is fixed, the upper moving with a velocity, either constant or oscillating to and fro at a range of frequencies.

The arrows in Figure 3 indicate the fluid velocities at different points in this shear flow. A shear rheometer will generate a small cell of material moving in this pattern and also measure the sliding forces on the plates necessary to maintain the flow. The 'shear stress' is this force per unit area of plate. Defined in this way, 'stress' becomes independent of the sample size, so it describes the internal properties of the material.

Since stress has, in principle, many components, corresponding to forces on the material surfaces both parallel to the flow and perpendicular to it, the rheology of soft matter can be richly complex. There is a historical and conceptual resonance with scattering experiments here: Karl Weissenberg invented one of the first rheometers for soft matter and called it a 'rheogoniometer' (reported in 1962), deliberately alluding to a moving detector device, or 'goniomenter' used in scattering experiments. As a former researcher on X-ray scattering from giant biomolecules, he conceived of rheology as a sort of scattering experiment—samples 'received' a range of geometries of strain and 'scattered' a range of geometries of stress out from its various surfaces, measured in an analogous way to detecting X-rays, light, or neutrons from a scattering experiment (as illustrated in Figure 13 on p. 52).

A sticky, stringy hot melt of a plastic, which flows as a liquid when trapped between the plates of a rheometer sliding slowly over each other, mimics the elasticity of rubber—it is viscoelastic—when the plates simply oscillate back and forth at high enough frequencies. Additionally, when a steady shearing flow rate on the same material is increased, its apparent viscosity reduces—it becomes a more runny liquid. This property, of great importance in processing for the plastics industry, is called 'shear-thinning'. We will delve into the molecular reasons for both in Chapter 3.

A range of colloid fluids—those composed of tiny suspended particles—exhibit an opposite effect when the particles are very densely packed. While they behave as fluids at low flow rates, they may display a strong *increase* in effective viscosity as the imposed flow rate increases, leading in some cases to a liquid-to-solid-like transition. Stirred cornflour paste is a familiar example—under slow stirring it flows easily, but a vigorous turn of the wooden spoon can cause the entire bowl to solidify momentarily. Such 'jamming transitions' have only very recently been explained theoretically in terms of changes in the fluids' microscopic structures (see Chapter 2).

Our introduction to soft matter science has already called on physics, chemistry, and engineering. More recent applications to the analysis of biological and bio-inspired phenomena increases the interdisciplinary palette even further. The consequences are as yet hard to predict, but already two promising directions for research have been generated. The first sheds new light on the physical basis of biological phenomena; the second draws inspiration from biology to define new research programmes in physics, chemistry, and materials engineering. An example is the rapidly growing field of 'active matter', the subject of this Very Short Introduction's last chapter, Chapter 6. The pickings for historians and philosophers of science interested in the social functioning of scientific communities will I suspect prove very rich when a little more dust (some of it 'active') has settled.

Each of the chapters that follow explores one of the soft matter family of materials in more depth. The experimental tools of microscopy, rheology, and scattering of light and other wave-like radiation will crop up as we investigate each one. Likewise, the common features of thermal (or 'Brownian') motion, intermediate (or 'mesoscopic') structures, slow dynamical processes, and universality together create a conceptual framework. Looking out for each of these features in the milky world of colloids, the stringy space of polymers, the foamy fluids of self-assembled soft matter, and the light-play of liquid crystals will help to connect their initially very different appearance and behaviour. It will also uncover some marvellously simple ways of thinking deeply about materials that, although their science might be new, have played important roles in human culture and technology for centuries, even millennia.

Chapter 2
Milkiness, muddiness, and inkiness

The ancient Akkadians discovered that the ability to write endowed them with the ability to govern, to account—in other words to rule. Babylonia arose in the third millennium BC on foundations of clay, not only as bricks but as the fired tablets into which the cuneiform characters of the first written language on Earth were pressed. History tends to focus on the remarkable invention of symbolic language, but the forgotten, humble yet wonderful technology of clay is just as essential. The paste that is clay before it is fired is a substance that flows under the press of a stylus, but unlike oil or water it does not flow back again, rather it retains its imprint for long enough for firing to make it permanent. Its rheology, to use the language of Chapter 1, is caught somehow between the states of solid and liquid; it is in many ways a technological marvel as well as a scientific mystery.

Later writing, in ancient Egypt, developed as marks on papyrus. In this case the new technology required was the idea of an *ink*—a coloured, opaque fluid that would flow from a quill or brush onto a surface, not spreading on contact but drying to leave a dark solid residue where the liquid had been. Of course we are now familiar with these useful properties of inks, but do we understand them? The first inks, those developed in Egypt, were made from mixing particles of soot into water. You can experience the marvel of the technology by trying this yourself—scrape a little soot deposited from a candle flame onto the underside of a pan into a glass of water. The result is disappointing, as the soot clumps together and sinks to the bottom. We will never know when and why someone around 3000 BC by the banks of the Nile decided to add a little gum-arabic to the water first. But that is the key move—now the tiny particles of carbon do two remarkable things. First, they stay separated, dispersing throughout the water. Second, they don't seem to sink to the bottom any more but remain uniformly suspended throughout the fluid—now qualifying as an ink.

The extraordinary phenomena of 'muddiness' (really an unusual set of rheological behaviours) and 'inkiness' (an optical property) are wrapped up in these examples with the unexpected stability they have in common. They share a technological history in the story of writing, but they possess another, microscopic, secret in common as well. They are both examples of 'colloids'—the fundamental class of soft matter constituted by dispersing very small particles of solid matter in a liquid environment. More surprising still is that, though technologically mastered for 5,000 years, most of their properties have been understood for little more than a century, and some for very much less than that. That understanding, in turn, has played a vital double role in opening widows of understanding into the microscopic world of atoms, molecules, and the larger structures of soft matter. For the colloidal state provided the final evidence that atoms existed, and now it provides to science a sort of 'modelling kit' that allows the construction of larger and slower versions of atoms and molecules, permitting direct visualization of the phenomena that lie hidden, in the case of the atoms themselves, at length scales a thousand times smaller. To follow this remarkable soft matter story, a good starting point is a visit to London's Royal Institution of the 1830s.

Michael Faraday on the 'Brownian movement'

The scientifically literate public of early Victorian London was clearly intrigued by the allure of unexplained phenomena.

Newspaper comment and public lectures on scientific topics were as fashionable as they were common. But in 1829 the place to go, and to be seen to go, for authoritative coverage of the most significant of the latest learning was the Royal Institution in Albemarle Street. Its Director Michael Faraday himself had begun the series of Friday evening discourses there that continue to this day. They became so popular that the London authorities were forced to solve an increasingly urgent traffic problem; the solution was the world's first 'one way street'. At 8 o'clock sharp the lecturer's bench at the focal point of the highly raked theatre draws the gaze of 300 pairs of eyes, and the expectant applause of as many pairs of hands, in a piece of scientific entertainment that one previous director calls 'the oldest stage in the West End'.

Faraday's public dissemination of science was not only entertaining but also topical and instructive, for on the evening of February 20th 1839, his choice of subject was the new mystery of 'the Brownian movement'. The polished public presentation of a new piece of research less than a year after its original publication was unusual. But Faraday clearly had a mission: as much to clear away misconceptions as to inform. His arguments are our clearest indication of the extraordinary public excitement that had followed Brown's first publication, not in the form of an article in a scientific journal as would be normal today but as a pamphlet published by its author. A botanist by original scientific interest, Brown had reported that particles within the translucent pollen grains of the wild flower *Clarkia pulchella* exhibited a continuous and random motion when observed under an optical microscope. He had gone on to investigate whether this was a unique property of these plant particles, or whether inorganic materials (such as stone, charcoal, or brick) when ground into suspensions of similar sized particles (5–10 microns) also possessed the microscopic motion. He found that any material, organic or inorganic, did so.

The summary lecture notes that Faraday prepared for his own use, as much as the record in the *Proceedings of the Royal*

Institution for that day, leave no doubt that he was determined not only to represent the content of Robert Brown's observations but also to employ a very particular rhetoric. 'Many think Mr. Brown has said things which he has not . . .' began his conclusion, 'all think that he must have meant this or that . . .'. Faraday devotes most of the lecture to making sure that his audience participates as fully as possible in the painstaking series of experiments that eliminated one by one the possible explanations of the 'oscillations', starting with his finding that they were most assuredly not evidence of a 'vital force' possessed only of living matter.

Brown had first covered the sample of water-suspended particles with a thin layer of oil or cleaved mica sheet to refute loss of fluid by evaporation as a possible cause. Not only that, but by so arranging samples that remained intact for days and weeks without drying out, the persistence of the motion over these much longer times could be checked. Perhaps small electrical charges on the particles were generating a network of tiny attractions and repulsions. Yet both Brown and Faraday knew that the presence of both positive and negative electric charge in matter that possesses overall neutrality must create another, slower, effect. When attached to mobile objects, any mutually attracting positive and negative charges eventually fall together, forming pairs and clusters of increasing numbers. Yet a fundamental characteristic of the Brownian motion is that time has no decaying effect on it whatsoever. In this spirit the Royal Institution's audience was treated to a lecture most remarkable for its series of negative results. No explanation within the imaginative possibilities of 1829 stood up to simple but brilliant observation under many carefully modified conditions.

Even the most thoroughgoing experimenter—and here Faraday clearly detects a kindred spirit—abhors the vacuum of total absence of explanation for their observations. The form in which he edited his remarks for the *Proceedings* must stand as one of the great passages of scientific prose:

> Naturally averse as Mr. Brown was to leaving the subject in this state, he, after many months investigation, published the discovery of these motions without pretending to account for the cause. Not asserting that a new power of matter was concerned, not denying that the powers with which we are acquainted might not be sufficient to originate the motion; but thinking it much more philosophical to acknowledge ignorance as to the mode of action, in these cases, and to suspend the judgement, than, by the assumption of an opinion, which must have been hypothetical, run the great risk of shackling the mind by the admission of error for truth.

This is a testament of huge respect—the pain of the absent explanation weighs in every word. Faraday admired Brown for his bare presentation the data and the eliminations of false explanation as far as he had been able to take them. Above all, Brown's resolute refusal to mislead the community into lazily assuming the existence of an approximate theory by casual concluding remarks draws Faraday's greatest approval.

Later, Faraday's own researches led him once more into the world of colloids. Work on the properties of very thin gold films by chemical washing of gold leaf created, as a by-product, beautifully coloured clear fluids. He guessed, correctly, that these were suspensions of particulate (colloidal) gold, and that their colouration was a result of the size, rather than the chemical makeup, of the particles. Gold, of course, is much denser than water, yet Faraday's gold colloids never settled out; in fact they can be inspected in his former basement laboratory in the Royal Institution to this day. Yet the enigmatic scribbled final line of his notes for that Friday evening discourse turn out to be the most prophetic: 'Connection with atomic or molecular philosophy…'

Einstein and the mathematics of molecular motion

By the end of the 19th century, several people had suggested that Brownian Motion could be a visible manifestation of the molecular graininess of matter—that the colloidal particles provided a bridge to the atomic world by being large enough to see (by way of optical microscopy) yet small enough to respond significantly to the buffetings of the molecules making up their suspending fluid. But it was Albert Einstein who, in one of his revolutionary papers of 1905, made the quantitative and conceptual leaps that allowed this insight to be developed into established science.

Einstein's paper, 'Über die von der molekularkinetischen Theorie der Wärme geforderte Bewegung von in ruhenden Flüssigkeiten suspendierten Teilchen' ('On the movement of particles suspended in fluids at rest as postulated by the molecular theory of heat'), takes two important steps in the theory of colloids, and of soft matter generally. First, he points out that the suspended particles, although typically of much larger dimension than molecules, are no different from them in their statistical properties—they are equally randomly distributed in space and, just like the molecules themselves, will also pick up random velocities from their thermal environment. Since the earlier work of James Clerk Maxwell and Boltzmann, it had been understood that the magnitude of the velocities of the independently moving particles, in a fluid or gas at thermal equilibrium, would correspond to their possessing an average kinetic energy proportional to the temperature of the fluid. In the case of a gas, this motion gives rise to the pressure exerted on the walls of its container—the huge number of collisions by thermally excited gas molecules against the wall average out to a constant force against each equal area. Likewise Einstein showed that suspended colloidal particles must act like a sort of gas, and would exert a gas-like pressure on a membrane

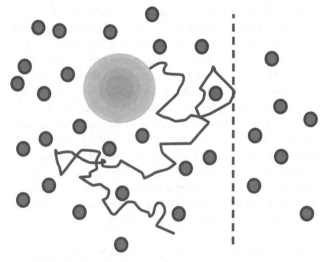

4. A visible colloidal particle executes a random motion as a consequence of its many collisions with invisible molecules of a suspending fluid. It contributes to a pressure-like force on a barrier (an 'osmotic membrane'—shown dashed) that it cannot pass through, but which is permeable to the molecules of the fluid.

that did not allow them to pass (see Figure 4). The mathematical form of this 'osmotic pressure' is *identical* to, and as simple as, that of Robert Boyle's celebrated law of ideal gases: in modern notation $p=nk_BT$. Here p is the pressure of a gas (or osmotic pressure of the colloids), n the number of particles per unit volume and T the absolute temperature in degrees Kelvin (K; on this scale the freezing point of water is 273 K and there are no negative temperatures). k_B is Boltzmann's constant, introduced in Chapter 1.

Einstein's second master-stroke was to write down an entirely new form of dynamical equation of motion, or rather to conceive a mathematical notation for a new type of force entering as a term in an equation of motion. He wanted to think about the colloid particles themselves as a complete system, without keeping track of each of the myriad solvent particles surrounding them.

He knew already that the solvent acted, at sufficiently large length scales, as a fluid, with a mean local velocity of flow and a resistance to the motion of objects through it proportional to its viscosity. What no-one had done before was to think in a similarly macroscopic way about the fluctuating nature of the forces that a small particle must suffer from a fluid that consisted of a finite, albeit very large, number of molecules. So Einstein wrote down a new sort of time-dependent force, a 'noise term' $\eta(t)$, whose actual course we do not know, but whose average properties we do. The consequence of an equation with an unknown term is that one cannot 'solve' it in the traditional way, but if instead one looks for the average over many particular trajectories, knowing the average properties of the unknown force can be enough.

Thinking deeply about the new equation for the motion of a colloidal particle in a fluid that exerted a continuous thermal noise on it led to three important consequences for its behaviour. First, the random buffeting on the particle would cause it to move through the fluid, also in a random manner known already as *diffusion*. Over time, the average displacement of the particle from its original position is as likely to be in one direction as another, but the average distance travelled grows as the square root of time. This is a very different motion from any steady velocity, where the distance grows proportionally with time. The difficulty that Brown, and experimenters after him, had encountered in measuring the velocity of the jittering motion of the suspended particles in their microscopes became clearer. Formally, diffusive motion does not even possess a well-defined velocity.

The second deep result of Einstein's theory of thermal noise was that the well-known viscous drag of colloidal particles through a fluid was intimately, and mathematically, related to the strength of the noisy force that caused them to diffuse. The stronger the noise, the greater the drag would be—both must, after all, come from the same summation of collisions with the surrounding, unresolved, molecules. This insight connected two apparently very different ways of looking at a complex thermal system: one can probe it using an external force, or simply measure the statistics and time-evolution of its inherent thermal fluctuations. In the example of a single colloidal particle, the first experiment corresponds to applying a force to it and measuring the speed of its subsequent drift; the second to recording its free diffusion without the application of any force. This turns out to be the simplest example of a very general set of correspondences called 'fluctuation–dissipation' relations. In all cases the greater the rate of diffusion in the second case, the lower the drag in the first.

The third result demonstrated how the colloidal state could act as a window onto the molecular world in a second sense, beyond the way in which the visible colloidal particles could model invisible molecules. For Einstein's explanation of Brownian motion depended explicitly on the existence of molecular matter—an indivisible continuum, as some leading scientists still maintained was the correct microstructural model of matter, would not generate the fluctuations in velocity and displacement. Furthermore, the effect could be quantified in such a way that the size of molecules could be pinned down. The smaller and more numerous they were, the more that a fluid would approach a smooth continuum, and the smaller the diffusion of Brownian motion would be.

Perrin's colloidal window onto the molecular world

An elegant experiment that drew on this third consequence of Einstein's 1905 paper was devised and reported by French physicist Charles Perrin in 1908. The idea drew on one of the questions that the existence of stable inks, or Faraday's gold colloids, elicited—why the suspended particles, denser than their supporting fluid, do not eventually sink to the bottom of the container. The answer within Einstein's theory was that the colloidal particles suspended in the background fluid sink in the same way that the Earth's atmosphere of nitrogen and oxygen 'sinks' onto the planet's surface. Recall that the thermodynamics of colloids is in no way different from gas molecules other than in the different scales of time and space. Seen from the Moon, the atmosphere has indeed 'sunk' onto the Earth's surface, but look closely and it has a finite height over which it thins gradually away. The reason is that, in equilibrium at a given temperature, the energy of motion acquired by a molecule can carry it to a characteristic height above the Earth's surface. It is straightforward to estimate this height. The mean thermal energy of motion at temperature T is just proportional to it, of magnitude $k_B T$, using the conversion factor of Bolzmann's constant. But the energy delivered to lifting an object against a force is easy to work out: if a molecule of mass m in a gravitational field of strength g is lifted a height h, the energy required is the product mgh. Equating these two energies gives a characteristic height of an atmosphere of $h = k_B T / mg$, which is a few tens of kilometres using a typical atmospheric temperature and the mass of nitrogen molecules. Now an 'atmosphere' of colloidal particles will be compressed over that of molecules of gas by the only term by which the expression for h can differ—the much greater mass m of the colloidal particles. If the size of the particles is a typical tenth of a micron, so a billion times the mass of an atom a thousand times smaller in diameter, the expected atmospheric height reduces to just 10 microns. Perrin's strategy was to peer down into such an atmosphere of colloidal particles with a microscope of very narrow depth of field—he would be able to discern only those particles within a very narrow range of a particular depth in suspension under the microscope (see Figure 5).

The experiment was a *tour de force*. Before he could make the key measurements themselves, Perrin had to devise a way of manufacturing suitable particles of nearly identical sizes. He found that a naturally occurring latex called gambose, often used as an additive to oil paints, would disperse into individual

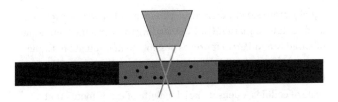

5. Schematic of Perrin's experiment: a microscope focuses narrowly on thin intervals of depth within a fluid cell containing a colloidal suspension at thermal equilibrium, allowing painstaking counting of the numbers of particles at each height.

spherical droplets in water. Many stages of separation by size, using repeated centrifuging of the suspensions, eventually resulted in sufficiently narrow distributions. Accurate measurements of diameters under the microscope were improved by lining up chains of particles. Finally, the painstaking work of counting particle numbers at different depths by eye by Perrin and his students, up to several thousand, provided statistics sufficient to test the Einstein prediction.

Not only did the experiment provide the final conclusive evidence for the reality of atoms and molecules by the detailed validation of Einstein's predictions, but it also delivered a quantitative measure of the scale of the atomic world. Working the formula for the colloidal atmosphere's height backwards from Perrin's measurement of it gave a value for Boltzmann's constant, the number k_B. One interpretation of k_B in the context of a gas, is as the 'pressure per molecule per degree of temperature', so combining Perrin's experiment with standard measurements of gases immediately allowed their pressures to be translated into measurements of numbers of molecules. Inky fluids had opened a window onto the molecular world.

Colloidal atoms: crystals, opals, and glasses

If colloidal particles, as Einstein and Perrin had showed, can do everything that molecules do but at a thousand times the size and equally more slowly, then they open up another way to explore the atomic state of matter. For by extending the synthetic techniques of Faraday and Perrin it proves possible to manipulate both the size and shape of colloidal particles and also the form of the forces between them. Modelling the forces between molecules, but at the scale of colloids, opens up a rich palette of experiments that explore the consequences for material properties of the sizes and shapes of the underlying forces.

Paradoxically, one of the most striking examples is the case of no force at all. A highly idealized, 'stripped-down' model for an atom or molecule is the 'hard sphere'. The label says everything we need to know about this highly idealized atomic model—such spheres only interact when they come into contact with each other, at which point they offer no compliance whatsoever. Although this seems a very unrealistic approach to understanding real materials, whose atoms always attract or repel in some way, it is nevertheless useful in separating the effects of pure mutually excluded volume from those arising from forces that act at longer ranges. By far the most significant consequence of molecular interactions for their fluids is the emergence of multiple states: gases, liquids, and solids. As real fluids are compressed from the gaseous state, there is typically a series of transitions (termed 'phase transitions') by which the material state changes suddenly at particular pressures and temperatures. First there is a condensation in which a proportion of the molecules adopt a higher density than the gas, remaining highly mobile and disorderedt but in close proximity to each other; this is the liquid phase. Compressing to higher densities still triggers the formation of another phase in which the molecules adopt a spatially rigid structure, also called a crystal. This is the solid phase.

An important question in the molecular science of matter is how these emergent series of phase transitions are connected to the local forces between the molecules. In particular, are the liquid and solid states dependent on the presence of attractive intermolecular interactions, or would they survive even if there were no attractive interaction at all? Common intuition says that both the liquid and solid states require some form of attractive interaction between particles. However, model calculations based on the 'hard sphere fluid', and its realization using spherical colloidal particles with carefully minimized interactions led to a surprising conclusion: while the existence of the liquid state does require short-range, attractive interactions to exist, spherical particles with no attraction at all will nevertheless crystallize above a critical density.

The possible existence of a hard sphere crystal phase had been suggested by the American pioneer of liquid state theory, John Kirkwood, in 1939. To understand how crystals might form with no apparent attraction requires a concept of the driver for any change of phase, plus an interesting twist of reason. Think of the boiling of a fluid. In spite of an attraction pulling the molecules of a fluid together—there is no force repelling them at all—at a sufficiently high temperature they form a phase, the gas phase, in which they are as far from each other as possible. The driver is *entropy*, which represents the degree of disorder: there are vastly more configurations available to the gas than the liquid (think about counting through all the places that each molecule in the gas might be compared to the few clustered around its neighbours in the liquid). Entropy is an account of these configurations. Providing that there is enough energy in the system to overcome the attractive forces—the more disordered state, the one with the greatest number of possible configurations and thus higher entropy—is the one realized.

A subtle version of entropy comparisons drives the hard sphere solid phase thanks to the packing efficiency of regular crystals. Spheres arranged on a carefully arrayed lattice (such as oranges on a greengrocer's stall) occupy less space than when they are randomly strewn. We may imagine two colloidal states—one in which each colloidal particle is free to move around a volume, or 'cage' centred on a point in a crystal lattice, and another in which the cage centres are arranged randomly. At first the two states are relatively sparse—spheres only rarely touch. This way of imagining the two structures, one ordered (the crystal) and one disordered (the gas), permits the counting of different possible configurations to be broken down, first, into a count of the possible cage centre placements and, second, into a count of the positions of the particles within their cages. So although the state of random placement of the cage centres possesses a greater number of possible configurations, as both states are compressed and the particle density increased, the spheres in the 'colloidal gas' state will have to touch permanently before those of the colloidal crystal. The better packing strategy of the lattice allows more compression than the gas. Furthermore, when the gas can be compressed no more, the freedom of its colloidal particles within their cages is reduced to zero, while those permitted to wander about cages centred on the lattice still find that they have room to do so—and many more possible configurations as a result. Within a range of concentrations approaching the maximal density of the closely packed crystal, an even better strategy to maximize the total number of configurations is for a fraction of the particles to adopt the crystal form and the remainder the (dense) gas. Fluid and crystal *coexist* in these conditions.

Early computer simulations of ideal hard spheres (in 1966) by William Hoover and Francis Ree identified the fraction of total volume filled with spheres (notated by the Greek symbol 'phi', ϕ) at which the crystal starts to appear as $\phi_f = 0.494$ and a higher filled volume at which the entire system would choose the crystalline packing as $\phi_s = 0.545$. Surprisingly, all this happens at considerably lower densities than the maximum available with random packing ('random close packing'), which is $\phi_{rcp} = 0.64$.

It would be twenty years before Dutch and British groups independently demonstrated experimentally the crystal phases of (nearly) hard sphere colloids. Careful synthesis of the colloidal particles, with exquisite control of their size distribution, was mandatory. Beyond that, the naturally occurring forces from electric charge and the mutual electric polarization of the particles needed to be minimized by carefully selected salts added to the solution. The process of concentrating the colloids needed equally precise control in order to provide time for the system to find its crystal phase before the colloids were jammed together into an immobile, but non-equilibrium, state. But when these conditions were fulfilled, objects of extraordinary beauty appeared, coruscating with colour.

The scattering of special wavelengths of X-rays, discovered by Max von Laue in 1914, was crucial in the elucidation of atomic crystal structures. Had the early X-ray films been sensitive to different wavelengths of X-rays, in the same way that colour film is to light, then the famous spotted and streaked early scattering pictures would have been highly polychrome. Colloidal particles are typically a thousand times the size of atoms, so their regular lattice planes selectively reflect, for precisely the same reason, radiation of a thousand times the wavelength of X-rays. This is just visible light. Figure 6 shows an electron micrograph revealing the spatial array of the colloidal crystal planes.

6. Colloidal crystals demonstrating crystal packing under the electron microscope.

If the iridescent appearance of synthetic colloidal crystals is reminiscent of the beautiful naturally occurring mineral opal, then that is because their characteristic colourations, that change with the angles of illumination and inspection, possess the same origin. The slowly cooled mineral forms regularly stacked planes of identical silica spheres typically about 0.1 micron in diameter. Opal is nature's colloidal crystal.

Features in flow of the soft and slow

Soft matter flows gently and sluggishly; our invitation to the colloidal state was the behaviour of colloidal fluids, inks, and pastes, whose flow properties were of their essence. Interesting rheology, as we saw in Chapter 1, arises from two of the characteristic properties of soft matter systems. First, their nanoscale structures offer new sources of forces and stresses within the fluids when they are deformed. Second, the slow timescales associated with these structures' relaxation back to their equilibrium state mean that commonplace flow rates created by stirring or pumping are able continually to hold the internal structures out of equilibrium.

The first experiment a soft matter scientist interested in rheology would do, faced with a structured fluid such as a colloid, is to measure the way that its viscosity changes with the most fundamental feature of its microscopic structure. In the case of a suspension of spherical colloidal particles without strong attractive interaction, the strongest variable to explore must be the particle concentration, or the fraction of the fluid volume occupied by the particles. At very low concentrations, we might expect the viscosity to remain close to that of the suspending fluid, since any pattern of flow will be hardly perturbed by the occasional colloid particle. However, near the concentrations at which the equilibrium phase would be a crystal, or close packed random, solid, much higher viscosities would be expected. In this regime large temporary clusters of particles need to squeeze and deform

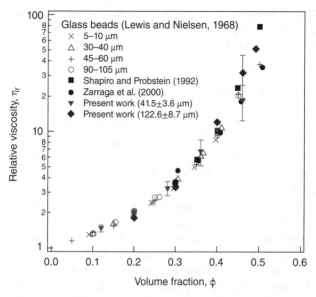

7. The measured viscosity of several particle suspensions in terms of the viscosity of the background fluid as the volume occupied by the particles varies.

past each other in order for the ensemble to flow—a process of much greater resistance.

The results of such experiments from various groups are plotted together in Figure 7. Immediately we notice another of the soft matter characteristics—that *universal* behaviour often emerges. Different sized hard spheres and different background fluids result in essentially the same remarkable upward curve of the viscosity as the particles pack closer together. The data is actually consistent with a divergence (a formal climb to infinity) of the viscosity, although it is somewhat of a surprise that this happens at rather lower volume fraction (about 0.5) rather than the close-packed solid value of 0.64. Changing the nature of the particles from hard to soft so that they are themselves able to deform on contact, or using electric charges to generate attractive or repulsive forces between the particles does generate different viscosity/concentration curves, although all have the general features of Figure 7.

Intuitively, this rapidly growing viscosity is familiar to us—it is reflected in the muds, cements, and sludges of children's sand-play, but it is not immediately clear where the huge stresses come from so far from actual close packing. We need to find another source of stress within the fluid than that arising purely from the motion of the solvent. A general rule in thermodynamics is that any change of a system away from equilibrium requires work to be done on it, that is, energy supplied from the environment. The simplest example is the pressure of gas in a cylinder: it possesses an equilibrated density; compression implies a change in that density, which in turn demands work to be done from outside (this is another way of thinking about pressure). But there are other ways in which the distribution of colloidal particles might be distorted from equilibrium. A shear flow, as described at the end of Chapter 1 (and see Figure 3) will do this as well, for example, although in a more subtle way (see Figure 8). For although, unlike a compression, the mean particle density is unchanged by this

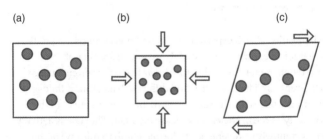

(a) (b) (c)

8. A local equilibrium distribution of colloidal particles in a suspension (a) may be distorted away from equilibrium by compression (b) or by shear (c). The large arrows indicate the Brownian stress forces that must be applied from its environment of (b) osmotic pressure and (c) shear stress.

flow, it does introduce special directions into the structure of the fluid. At equilibrium, the mean distances from a given particle to its neighbours is isotropic—independent of direction. But after a shear, the pattern of these near-neighbours is distorted: in the quadrants of the flow field where fluid is moving away, nearest neighbours will tend to be a little further away, and in the other quadrants a little nearer, than at equilibrium. The structure has been distorted away from its equilibrium average by the flow—so that must have required external work. That work, in turn, must have been done against a stress arising purely from the spatial distribution of the colloids.

The thermodynamic, or Brownian, shear stress is really just the equivalent of osmotic pressure for the shear geometry of distortion. Recognizing that these stresses arise from the distortion of spatially extended patterns of particle structure suggests a characteristic by which they can be experimentally recognized as distinct from the shear stress of background fluid viscosity. For if a flow is stopped, purely viscous stresses vanish immediately—they are dependent on the shear rate, so as soon as that falls to zero, viscous stresses disappear. But the Brownian colloidal stresses have a different origin in the spatial arrangements of particles. Those take some time to return to equilibrium, because particles need to diffuse across distances of order of their mean spacing to restore the original, and isotropic, distribution. A stress that decays with time we call viscoelastic. The viscoelastic stresses in colloidal fluids are rather subtle in most cases—we will encounter much stronger examples in the case of polymers, but at high densities these are the stresses by which the slow flow viscosity diverges. Experiments observing stress relaxation in colloidal dispersions over time, or in characteristic response to oscillating shear flows at different frequencies were first performed by researchers at Twente in the Netherlands in the 1980s and shown to be in agreement with Brownian stresses relaxing by particle diffusion in calculations by John Brady at Caltech.

The corn starch experiment—fast flows and jamming

Of all the soft-matter-related experiments that make it onto YouTube, perhaps the most spectacular presents a giant sized version of a simple but remarkable kitchen phenomenon.
A dispersion of corn flour in water or milk is a colloidal fluid—the corn starch granules are less than a micron in diameter. At high particle density, with just a little fluid to suspend them below the packing concentration, the milky fluid can be stirred or poured very easily. But, as we noted in Chapter 1, if the spatula is moved suddenly, and under high force, the corn flour liquid becomes momentarily solid. Only when the frantic stirrer releases the force does the suspension refluidize. The YouTube spectacular replaces the mixing bowl with a 5-metre-long tank, and the spatula with a line of students or a TV presenter who stride out onto the surface of the liquid.[1] Although the viewers are given a demonstration of its low viscosity when gently stirred, the rapid imposition of stress from the weight of the walkers creates the same transitory effective solid. Providing they do not lose faith and hesitate, as St Peter did in a similar situation (Matthew 14:28–31), they do not sink...and arrive dry at the other end.

Surprisingly, the underlying cause of this phenomenon of 'discontinuous shear hardening' has only been elucidated very recently, through a combination of simulation, experiment, and theoretical physics. For a long time the community was diverted by a correct, but misleading, result in the fluid dynamics of hard and smooth particles forced together towards close contact. As two particles in a fluid approach each other closely, the gap between them becomes ever narrower. As a result, the pressure forces needed to drive the remaining fluid out of the narrowing gap grow ever higher. It turns out that due to this pressure

[1] An amusing version can be sourced at: https://www.youtube.com/watch?v=JkSIymQ73oc.

growth, formally speaking, perfectly smooth particles can never touch. But this means that there must always be a lubricating layer of fluid between them, which prevents any emergence of the solid-like transition under flow of corn starch.

Yet dry powders, and models of them, do exhibit such 'jamming' transitions. Figure 9 indicates why in a beautiful visualization of a two-dimensional particle flow experiment.

As the shear flow brings particles into contact, so they form structures of compressed chains and networks, oriented along the direction of net local compression in the flow. The frictional nature of the surface contacts ensure that, for a period of time at least, a force chain survives before buckling and collapsing, but long enough for others to form. A transient solid results. Fluid

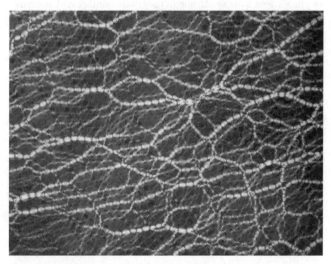

9. A visualization of force chains in a strongly sheared quasi-two-dimensional granular system. Photo-elastic discs under polarized light show 'jammed' chains of particles transmitting force building up in resistance to the shear flow.

suspensions under flow can allow surface friction to 'bite', in spite of the theorem of no approach. The reason is that real colloidal particles are not perfectly smooth. Asperities on their surfaces may begin to interlock before the lubrication flow problem starts to prevent further approach. Matthieu Wyart in Strasbourg and Michael Cates in Cambridge have shown that the rapid shear induced jumps from liquid-like to solid-like states in suspensions of frictional particles emerge naturally from a flow response that contains both the viscosity increase of close-approach and the parallel possibility of forming stress chains. The experimental picture is still developing, with the recent deployment of Magnetic Resonance Imaging (MRI) imaging able to track local regions of fluid and solid.

Even the simplest possible soft matter system, in which the nanoscale objects are spherical particles, causes a rich collection of phenomena to arise. Although they make contact with everyday experience, many are only now at the forefront of research questions, such as the strange fluid dynamics of sedimentation and the properties of colloidal liquid and glassy phases. For now, however, we will take our leave of the milky colloids and turn to a very different class of soft matter—and the molecular origin of 'stickiness'.

Chapter 3
Sliminess and stickiness

In the spring of 1920, a young, high-flying professor of organic chemistry at the Technical University of Zürich published a paper, 'Über Polymerisation', that coined both a new word and advanced a shocking new idea. With a background in botany, Hermann Staudinger had decided that he ought also to take up chemistry, foreseeing that the future of understanding living matter would be at the molecular, and chemical, level. He proved extremely able within the burgeoning field of synthesizing new compounds of carbon—the entire chemical division known as 'organic chemistry'. The enormous, profligate, superabundance of chemical possibility that the carbon atom offers is quite unique among elements; without it life would not be possible. It also proved the key to the replacement of many natural products in short supply with artificial ones in Germany during World War I. For example, the young chemist had found, during those years, a route to an artificial coffee aroma. Two of the organic molecules Staudinger had found good routes to synthesize contained 'double bonds'— one of the key features of carbon chemistry that enables its electrons to be shared with other atoms in compounds in many different ways.

Butadiene and isoprene have similar structures—a short 'backbone' of four carbon atoms with double bonds at the end (see Figure 10). In the case of isoprene, one of the hydrogen atoms of

(a)   (b)

10. Structural formulae for the reactive organic molecules (a) butadiene and (b) isoprene. The reactive 'double bonds' are shown with double lines.

butadiene is replaced by a 'methyl' (CH_3) group. Staudinger's experimental reactions using these molecules sometimes resulted in a solid precipitate whose molecular weights (the weight of a molecule of a compound in units where a hydrogen atom is 1 and a carbon 12) were not measurable. This was, by itself, not a surprise, as the double bonds could provide attractive forces able to aggregate organic molecules that contained them. The real puzzle emerged when the double bonds were removed by adding hydrogen ('hydrogenation' reactions were well known), for the aggregates stubbornly refused to break up. That was the point at which Staudinger broke away from his community by making a wild and imaginative suggestion. He proposed that the original aggregates were not temporary conglomerates at all but giant linear molecules generated by the prior opening up of the double bonds, so that the small molecules could 'chemically hold hands' and string out indefinitely into the monsters he termed 'macromolecules'. No one believed him. Furthermore, to a chemical community that prided themselves on exactitude and rigorous measurement, the very notion of a molecule of indeterminate molecular weight was appalling and unacceptable. Staudinger had a fight on his hands. Among the kindest of comments was from Heinrich Wieland, winner of the 1927 Nobel Prize in Chemistry, who advised, 'Dear colleague, drop the idea of large molecules; organic molecules with a molecular weight higher than 5000 do not exist. Purify your products, such as rubber, then they will crystallize and prove to be low molecular compounds!'

The paradox was that macromolecules had been employed by humans since before written records. There is archaeological evidence that several ancient peoples of Central America used natural rubber, found in the latex of the indigenous Hevea tree. Charles Goodyear had, in 1844, patented his vulcanization process, which could stabilize the slowly flowing rubber into a solid, and the automotive industry was using rubber tyres long before Staudinger made his proposal. The main long-chain molecule present in natural rubber is, remarkably, a form of the polyisoprene that he was convinced he had synthesized.

Familiarity is not the same thing as understanding—it is one of the aims of science to distinguish the two. We may now, as users of materials, be used to the extreme extendibility of rubber, as were Victorian users of natural rubber, or familiar with the immensely sluggish and rubbery flow of latex, but that does not mean that we understand them, or have appreciated what extraordinary properties they possess. Every other solid typically breaks when it has been deformed by less than 10 per cent of its length, yet rubbers can stretch over 100 per cent more than their original length repeatedly before they resist further extension, and still without brittle failure. Understanding these unorthodox properties of the polymer liquid and gel state turned out to be one of the first successes of soft matter science. The seed was sown early. One of the experiments that convinced Staudinger that he had made his 'macromolecule' was made possible when a solvent was found for the isoprene product that others claimed had to be a clump of the 'colloidal' matter we met in Chapter 2. The resulting syrupy fluid possessed a viscosity, at low concentration of the dissolved material, orders of magnitude more than could be explained by dissolving a small number of small molecules into the easily flowing organic fluids. It seemed that the new molecules were taking up much more of the space in the fluid than they ought to be.

How big is a polymer molecule?

Sam Edwards had fallen in love with mathematical physics as a young student in Cambridge and had gone to Harvard University to undertake a PhD with Julian Schwinger, who had just shared a Nobel Prize for creating a workable quantum theory of light. Returning to the UK for a post in physics at the University of Manchester, and seeking new territory for physics, he imaginatively called in one day at the chemistry department next door, inviting an organic chemist, Geoffrey Gee, to talk through current challenges. Gee mentioned Staudinger's challenge: how much space would a polymer molecule occupy when in solution? Edwards soon realized that many of the mathematical techniques that had been developed to deal with the quantum mechanics of electrons, photons (the quantum particles of light) and their interaction in solids, could be translated into tools for solving polymer problems such as this one. With this insight from the unexpected quarter of the 'quantum field theory' of light, polymer chemistry gave birth to polymer physics.

To see how this works, a simple but profound mathematical picture from one of Schwinger's co-Laureates is required. Richard Feynman's early work on quantum theory had provided a third way of thinking about this new physics of matter. The 1920s had seen two previous approaches in the 'wave mechanics' of Erwin Schrödinger and the 'matrix mechanics' of Werner Heisenberg. Initially appearing contradictory, it was the British physicist Paul Dirac who showed that they were really two languages for the same idea. Feynman added a third language—and moreover one that could be represented in picture form. A classical particle in motion, he reasoned, moves in a well-defined path from starting point to finish. Throw a ball to your friend, and you can trace its trajectory. A quantum particle moves in an entirely different and strange way: in some, at least mathematical, sense it travels along all possible paths connecting the starting point A and the end

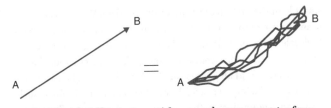

11. The full probability that a particle, or a polymer propagates from A to B (represented by the arrow on left), can be calculated as a weighted sum over all possible paths (on right). A situation close to the classical limit (for the quantum particle) or high-tension limit (for the polymer) is illustrated, where the highly probable paths cluster around a single trajectory.

point B. The probability that it arrives at the end point (nothing in quantum mechanics is certain) is then calculated from a sum over all these possible paths, each one contributing a particular number to the sum. The mathematics is shown in picture form in Figure 11.

The picture gives away the connection to the polymer problem. Suppose that one end of a polymer molecule starts at A; one way of getting at the problem of its overall size is to ask about the probability that the other end arrives at another point B. The polymer is not a quantum object, but the sum over paths arises in a similar way, in this case because it is a *statistical* object. A real polymer chain will be in rapid motion, due to the Brownian fluctuations of the solvent, so it will explore all possible states and configurations in the solution. Furthermore, any experimental measurement of the solution would average over many individual polymer chains. All would depend on the average property, and that would result from a sum over paths. The formal similarity between the quantum mechanics of a particle and the 'statistical mechanics' of a polymer lies at the heart of soft matter physics.

To go a little further into this deep conceptual mapping between the quantum particles and polymers, the length of the polymer chain corresponds (though a mathematical twist) to the time variable in the motion of the quantum particle. Varying the temperature of the polymer system can be compared with varying the scale of the particle system between quantum and classical limits. For example, extreme tension applied to a polymer chain stretches it so that all fluctuations in its typical path lie close to a 'classical' trajectory of a straight line between its endpoints. Such a situation is equivalent to the physics of the large-mass limit in the quantum case, where this is the dominant classical trajectory of the particle. There is a profound aesthetic in such conceptual links between ostensibly distinct worlds of physics.

An important conceptual step has been taken here in moving from a chemical description of a polymer, in which the carbon and other atoms in each group are accounted for, to a physical description of a 'stripped down' flexible path through space. This simplification is none other than the classic soft matter property of coarse-graining, to length scales larger than the atomic, that we met in Chapter 1. The reason that a faithful physical picture of polymers can be coarse-grained in this way is that even small amounts of flexibility at the molecular scale builds up to the same, complete flexibility at sufficiently large scales. The isoprene and butadiene 'monomer' building blocks of Figure 10, when chemically polymerized into a chain, are able to rotate relative to each other along their carbon-carbon bonds. Although this gives only relatively unconstrained movement over a few monomers, longer pieces of chain may adopt paths that point in any direction. This is the first key step in solving Gee and Edwards' problem of the size of a polymer chain in solution. For the completely unconstrained chain, at the longest length scales when it can be assumed to be completely flexible, takes the familiar form of a 'random walk'. This is another common object in physics: the path of a diffusing particle taking random steps as it collides with the molecules in a gas or a liquid, the conduction of heat through an insulator, or the way measurement errors add as they are combined—these are all examples, along with the polymer, of a quantity that grows by the sum of steps of fixed size but random direction.

To understand the properties of random walks, it is important to ask the right question. They are random, statistical objects, so the appropriate measures for them must be averages. But asking what the average displacement is from one end of a random walk to the other does not help. It is as likely to move to the right as to the left, so the mean must be exactly zero. Much more helpful is to consider the average *square* of the displacement, for this is by definition always positive. A wonderfully general and applicable result is that this quantity grows proportionally with the number of steps in the walk. Alternatively, the mean (positive) distance from one end of a random walk to the other increases as the *square root* of the number of steps. Note that this is a far less effective way of moving away from a starting point than travelling in a straight path. In such 'ballistic' motion a moving object quadruples its distance from its starting point after four times the time taken. In the diffusive, random case the mean distance has only doubled, and after nine times the time taken, only tripled. The spatial equivalent of the ballistic motion of a particle is a straight rod; the spatial equivalent of diffusive motion is the polymer. A simulation of such a random walk is shown in Figure 12.

The structure of the random walk is, however, far more open than that of a completely dense molecule. This provides the clue, almost as Staudinger suspected, to the high viscosities of polymer solutions. The effect on the viscosity arises from the amount of volume 'spanned' or taken up by the chains. Were these collapsed into colloidal lumps, they would occupy a much smaller space (see the illustrated collapsed sphere in Figure 12). Staudinger himself believed that polymers would persist with rod-like, linear configurations, so reaching out even further into the solution. The realization that even small amounts of local flexibility would eventually lead to the random walk structure emerged in the 1930s and 1940s from the German and American physical chemists Werner Kuhn and Paul Flory.

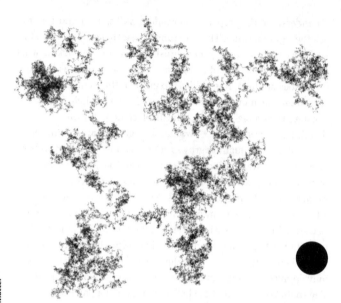

12. A simulated random walk of two million steps projected onto two dimensions so that overlapping points are shown darker. If the chain were collapsed into a dense particle it would occupy a sphere of the size shown.

There is one more, vital, subtlety hidden within the problem of polymer configuration and size. A careful look at the large simulated random walk of Figure 41 in Chapter 6 will reveal that there are many places where the walk may be attempting to cross itself. Although a diffusing particle executing a random walk may naturally retrace over a previous step, a material embodiment of a walk, as a permanent string, may not. It was with this problem of 'excluded volume' that Sam Edwards principally engaged with on leaving Henry Gee's office in Manchester. Feynman, Schwinger, and Sin-Itiro Tomonaga had had to tackle a similar problem in what they termed the 'self-energy' of an electron, and the mathematical tools for counting such interactions were Edwards' inspiration for working out how an otherwise random polymer chain in a solution would avoid itself. Flory and Edwards both arrived at a conjecture, later established by field theory techniques evolved in the 1970s, that very large chains would 'swell' under their own excluded volume in an identical way when viewed at all length scales. The mathematics of the result is so simple it is worth recording in a single line:

An ideal random walk of N steps each of length b moves an average distance from its origin of R, such that $R \cong bN^{1/2}$ (the square-root law of diffusion); a self-avoiding walk modifies this to the slightly larger structure of $R \cong bN^{3/5}$.

These simple formulas illustrate beautifully the central ingredient of soft matter systems—that they support structure at a length scale much larger than that of the atom, but still small enough to be microscopic, and subject to the effect of Brownian motion and thermodynamic equilibrium. The atomic length scale, that of the chemical monomers of Figure 10, is the small length b. The size of the much larger polymer molecule is the R—they are separated by the multiple of the large number N of links in the chain (albeit raised to a fractional power). Polymer chains of a million links are not unusual—in that case the simple random walk model (using $R \cong bN^{1/2}$) would predict a polymer molecular span a thousand times larger than the monomer. The swollen, self-excluded, chain (using $R \cong bN^{3/5}$) gives an even larger factor of 4,000. The very reason that R is meaningful as an average object is that the polymer coils are, when in solution, in constant motion, adopting configurations like Figure 41 of Chapter 6 momentarily, then coiling around into another, similar, shape, then another.

Random rubber

For all that these considerations of physics ideas from quantum mechanics and the mathematics of random walks might appear arcane, they provide the physical insight, as well as computational strategies, into practical understanding of important and everyday phenomena such as rubber elasticity and toughness of reinforced plastics. Rubber is the material created when many long polymer chain molecules are linked into a giant lattice. We can now begin to understand its peculiar properties, and in particular how it is so easily and so robustly extendable. Rubber is a strange solid because, at the molecular length scale, the local structure and dynamics are indistinguishable from that of a liquid: local sections of polymer chains are in continual thermal motion around each other just as they would be in a concentrated solution. The solid order appears only at the far larger scale of the distances between the chemical cross-links that tie the chains together at junction points. These cross-links fluctuate in their exact place but only around fixed mean positions. When a rubber band is stretched, the force with which it responds comes not from any physics at the atomistic level, such as the stretching of molecular bonds; rather, it is an emergent effect arising from the increased restriction on the multiplicity of possible paths of all the polymer chains belonging to the rubber network. As the cross-links that mark their end points are pulled apart, so the chains between them are required to stretch out (adopting the stretched configurations depicted in Figure 11, rather than the random walk of Figure 12). There are simply fewer paths by which a chain may link two distant points than two close ones. Pulling a polymer chain into a state in which there are fewer realizations needs work, in just the same way that compressing a gas does. There are fewer ways of arranging gas molecules in a small volume than there are in a large one (recall from Chapter 2 that we say that in such cases of the smaller volume for the gas, or the more stretched configuration for the polymer chain, that the entropy is lower). Each polymer chain acts, in consequence, as a tiny thermal 'spring'. Their minute individual forces when stretched have now been detected through the technique of 'atomic force microscopy', by which a tiny silicon stylus picks up and pulls on a single chain attached to a surface.

Again, a simple equation, very similar to one we met in Chapter 2, illustrates this new link between the physics of rubber and the physics of a gas. We saw in Chapter 2 that the ideal gas law for the pressure p exerted by n gas molecules in a volume V is given by:

$$pV = nk_BT$$

The pressure depends only on the number of molecules per unit volume and on the absolute temperature T (with the appearance of Boltzmann's constant k_B as a 'currency converter'). The result for the elastic pressure (this is a pulling force, not a pushing force as with the gas) p_e from a rubber containing n cross-links in a volume V is:

$$p_eV = nk_BT$$

The structures are precisely the same. Cross-links in a piece of rubber act as a sort of negative elastic gas. Pulling on an elastic band couples our fingers directly to the Brownian motion of millions of macromolecules. There is one more very tactile way in which anyone can test this analogy. For when a gas is compressed, such as in pumping up a bicycle tyre, it gets hot (try touching the metal end piece of the pump immediately after the task). The only place for the energy of work in compressing the air to go is in heating it up. If stretching a rubber is indeed analogous to compressing a gas, then it should also warm up under the process. The easiest way to detect this is through the very temperature-sensitive skin on our lower lip: pull hard on a thick rubber band as taut as it will go, and immediately hold it in that state against your lip. Rapid relaxation of the rubber will conversely cool it—in some ways an even more remarkable feeling. This simple experiment itself has an interesting history. Its discovery is due to the Cumbrian Quaker scholar and scientist John Gough, acquaintance of John Dalton and author of a remarkable 1803 paper on the properties of rubber. Gough had lost his sight, from polio, at the age of 3; his scientific work concentrated on phenomena that are recorded best through other senses.

Detecting the dance of the polymer world

Experimental verification of the theories for polymer coil sizes and configurations came indirectly at first, as we have seen, through their effect in modifying the viscosity of solutions. Direct microscopy of polymers was not possible (and is still only available through first flash-freezing the fluids before electron microscopy), but direct structural information could be extracted through careful measurement of light passed through their solutions, using ideas first worked out in the context of atmospheric scattering by Lord Rayleigh in the 1870s. Ideally, laser-produced light is sent into an optically clear cell containing the polymer, while detectors receive the light scattered at all angles (see Figure 13). The coil size of the largest polymers in solution are not very much smaller than the wavelengths of visible light, so light scattered from

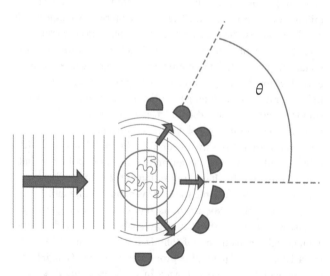

13. A schematic of a scattering experiment: incoming (coherent) waves from the left encounter the sample and are scattered and detected at all scattering angles θ.

different parts of the chains interfere slightly, giving a characteristic angular dependence to the outgoing intensity.

The interference of light scattered by different polymer chains carries additional information on the pattern of their density and on any repulsive or attractive interactions they encounter when they begin to overlap. As with all such scattering experiments, information contained at large scattering angles corresponds to the smaller structures in the sample, while small angles close to the direction of the incoming radiation encode the largest structures. At these largest length scales in polymer solutions, fluctuations in the overall polymer density shape the outgoing light.

In so far as soft matter possesses large structures, as we saw in Chapter 1, it also typically generates slow dynamics (all relative to the atomic world). The scattering of light proved to be an introduction to the world of polymer dynamics as well, for it is not difficult to resolve the fluctuations in time of the scattered light, which in turn betray how fast the polymer coils are moving and changing their configurations. The principal motion detectable by the wavelengths of visible light is the diffusion of entire polymer chains. Here we meet with the random walk again—but now the dance is in time rather than in space. The random walks set into the chemical space of the polymers themselves now execute random walks in time. For the same reason that the overall size of chains grows as the square root of their chemical path lengths, so does the distance diffused by the coils over an experiment. Removing the electronic detectors from an experiment such as in Figure 13, and replacing them with a sheet of paper, can give a stark and beautiful illustration of this scaling law of space with time. For at any one moment, the pattern of light scattered to the side of the main beam contains patches, or 'speckles' of light (see Figure 14). The bright places occur where there is a temporary net constructive interference of waves of light from different parts of the sample, the dark spaces where the interference is destructive.

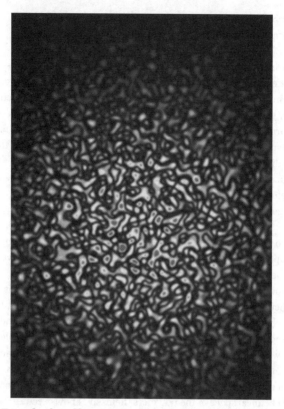

14. Example of speckle pattern—a snapshot from a light scattering experiment.

As the polymer chains diffuse about, different places will assume constructive and destructive interference, and so the speckles will twinkle, appearing and disappearing, while maintaining the same statistical distribution of shapes and scales. The rate at which they do so follows the typical timescale for structural change at the spatial scale that dominates the scattering. So at large angles to the side of the beam, the speckles dance and flicker rapidly, while at small angles close to the beam, when interference comes from much larger structures in the sample, they appear and drift more slowly.

The twisting and coiling of structures inside the chains themselves is too small to be resolved by light, but fortunately quantum mechanics comes to the aid of experimental soft matter physics just as helpfully as it did to its theory. For atomic particles, such as protons, neutrons, and electrons, also assume wave-like properties, so they scatter in a similar way to light from objects with which they interact. A particularly useful technique in investigating the structure and dynamics of soft matter is the scattering of neutrons. As neutral particles, they are not interrupted by the electron clouds filling most of the space in materials but only scatter from the atomic nuclei present. The most strongly scattering nucleus is that of hydrogen, which is dominant in the organic chemistry underlying soft matter. Furthermore, the wavelength of neutrons from the sources used for scientific work can be moderated to the scale of a nanometre or less, which means that in a single scattering experiment through small and wide angles, length scales from the atomic up to a few tens of nanometres can be probed simultaneously— exactly the typical size-range of structures inside a polymer coil. The equivalent of the 'speckle pattern' experiment can be carried out as well, measuring the typical dynamics on each length scale.

The result from the first neutron experiments on polymers, conducted in the early 1970s, not only confirmed the random walk structures for the entire chains that had been proposed theoretically and measured indirectly before but were also able to probe inside the chains themselves to their inner structure. The neutrons unveiled another type of universality: looking again at the random walk portrayed in Figure 12, it is hard to find any particular size scale in it that stands out. Any small piece of the random walk, on average, looks like a miniature copy of the entire chain, and small pieces of that sub-section likewise. A polymer is an example of the class of objects known as 'fractals'. They have a

15. **A regular fractal: 'Sierpeinski's gasket' created by repeated removal of the central inverted equilateral triangle from one with sides twice as long.**

special sort of symmetry, very different to that of geometric solids like the cube or tetrahedron that are unchanged (symmetric) under specific rotations, or to lattices that are symmetric under translations by whole numbers of repeats. Instead, fractals are symmetric under *change of scale*.

Figure 15 gives a famous example of a regular fractal—each porous equilateral triangle is an exact, but scaled, copy of the whole. Polymer coils are similarly fractal, but in a random, or statistical, sense: each sub-chain is statistically a random walk with the same properties as the entire chain. The pattern of scattered neutrons from fractal polymers is reflective of this scale-invariant symmetry. Each change of scale (detected by a change of scale in the scattering angle) corresponds to a constant factor reduction in the intensity detected. In other words, the scattering pattern is also a fractal.

The structural self-similarity across length scales generates a similar pattern in the dynamical behaviour of polymers. The noisy fluctuations of similar patterns in small and large sub-chains will themselves be similar, except for a change in their characteristic rates. After an elastic deformation of a polymer chain, for instance, local pieces randomize once more very quickly, followed by larger sections of the chain composed of those local pieces, then by much larger sections still, and so on in a cascade upwards in length and time until the entire chain has regained equilibrium. The speckle signal from such a sequence of events is self-similar, or fractal, in time as well as in space, reflecting the same connection of space and timescales of the polymer chains themselves.

Entangling elastic liquids

There is a fluid effect of polymers that the simple physical ideas of this chapter have yet to explain, though we have been building towards it in the chemistry and physics of first single chains and then the cross-linked rubber. This is the sticky sliminess of rubbery liquids when they are not cross-linked. Rubber latex is a strange liquid, for it flows, yet also forms strings and shows much of the elasticity that it will possess permanently when vulcanized (cross-linked into a network). This same simultaneous property of elasticity and fluidity is exhibited by plastic in its melted form—a fluid of densely packed polymer molecules.

A clue is found in one other fundamental property of polymers that is required to understand the concentrated state of their fluids. String-like structures cannot simply flow unrestrictedly like the much simpler and smaller molecules of simple liquids like water or oil, for the strings would have to pass through each other to do so, a dynamical move only possible if they are cut and reformed after the crossing event. (Wonderfully, there exist proteins in bacteria that perform precisely this process on the bacterial DNA when it needs to disentangle itself into the two 'daughter' cells during cell division.) We say that the polymers possess *topological* interactions. Topology is the study of those aspects of form that remain irrespective of stretching and deformation (but not cutting or joining). A famous example is the number of holes in an object—a teacup is topologically equivalent to a doughnut, since both contain just one hole and could be deformed continuously into each other. Similarly, polymer strings may change their configurations, but if they are entangled with each other in a particular way, just stretching them will not change that. At first sight, such a massively complex system of entanglement between many polymer chains would seem to be forever beyond comprehension or quantitative calculation.

An imaginative and simple way to think about the problem of polymer chain motion in the entangled state was conceived by the French physicist (and later Nobel Laurate), Pierre-Gilles de Gennes, in 1972. He was thinking about a weak solid of rubbery networks, but supposed one molecule *not* to be attached, but instead free to wander through the three-dimensional random latticework of the other linked chains. Then there would be, for this special string molecule, only one way to disentangle itself from its original neighbours, which is to slide along its own contour—the tortuous path created by the neighbouring strings and their cross-links at every turn. To invoke a culinary analogy, this is the very motion that we generate when we eat noodles or spaghetti politely, winding up a few at a time with a fork—this makes them slither past their neighbours and out of the tangle. In the case of polymers, the slithering motion would not require the force of a fork, for the internal Brownian motion of the polymer would generate it, albeit as a sequence of random steps one way then another.

De Gennes's conception of the slithering Brownian motion of a free polymer chain in a network reminded him of a snake

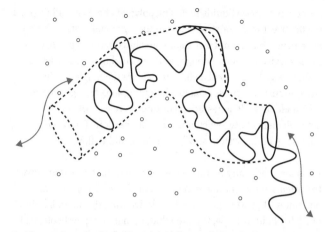

16. The meshwork of neighbouring polymer chains (represented as dots in the diagram) surrounding a given chain (solid curve) creates an effective tube-like constraint around it (dashed line) so that its random Brownian motion is confined to a slithering back and forth along the tube, indicated by the double arrows.

slithering in reptilian fashion along a tunnel or tube, inspiring the coining of a new word for this special dynamic mode, which he called 'reptation'. The idea is illustrated in Figure 16.

Further into the same decade, Sam Edwards and a chemical engineer Masao Doi, working together at the Cavendish Laboratory in Cambridge, realized that the same imaginative key—to think about one molecule at a time, constrained in its motion by its many neighbours—would also unlock the problem of the polymer melt, even when no molecule is special and where no permanent cross-links tie the stringy structure together. Each can only move along its own contour even when all other molecules are also in motion, and only when each molecule has crawled out of its straitjacket of entangling constraints—fortunately they may all do this simultaneously—will the composite material be able to flow a small amount. So repeated escape from the entangled structures and the re-creation of new entanglements is the molecular picture for how polymeric liquids flow. The polymer chains are all trapped in effective 'tubes' constructed from the entanglements with their near-neighbours, that trace their convoluted paths. The effective tubes survive just long enough, as those neighbours themselves slither away, to allow each chain to explore the tangled stringy surroundings by this special random toing-and-froing along their self-made tubes. The resulting continuous release of old entanglements between chains, and the creation of new ones, allows the entire system slowly to deform and to flow.

It's an extraordinary picture—but more than that, one that powers the calculation of numbers in the prediction of experimental outcomes. It proved quick work to derive what the 'memory' time of such a fluid must be, the time that, in macroscopic rheological experiments, divides the elastic behaviour of the melt at short times from the fluid flow at longer times. Nothing can deform or flow very much while all the polymer chains are still trapped by most of their original set of entanglements, but when all have been shunned, and each molecule has explored a new space within the liquid, some flow is set free. So the memory time is just the duration of escape of an average polymer chain from its original tube. This can indeed become very protracted as we imagine experiments with increasingly longer chains. Not only must each longer chain wander back and forth at random along its confining tube for longer distances, but each step is also slower in proportion to the number of units it must drag along. Everything adds up to a tube-escape time that increases as the *cube* of the length of the polymer chains—double the length, and the emergent memory time increases eightfold. When this prediction first came to light, no new experiments were needed to check it, for the result lay very close to data that had long been known to polymer scientists in universities and in the polymer industry alike.

The dominance of topological interactions in the polymer melt state is the principal cause of the long timescales in the fluids. Its rationalization with the tube picture has proved equally powerful in exploring other aspects of the entangled state. In strong flows, for example, the chains can align and begin to disentangle, much like a microscopic version of combing hair. In this state the effective viscosity is much lower. A way to increase the viscosity, on the other hand, is to play on the crucial, entanglement-creating topology (recall the chain properties that cannot be changed just by deformation) and to alter the character of the polymer chains themselves by branching them into star-, tree-, or comb-shaped architectures. Each different topology of molecule exhibits a different rheological response under flow. When every molecule is a many-armed comb, for example, the side arms are responsible for the relaxation of rubbery stress at short times, while the long backbones relax very much more slowly. Understanding this level of physics is now enabling industry to design new plastics at the molecular level, with specific material properties in mind.

At the time of writing, one particular architecture of polymer is attracting a lot of attention because of the permanence of its topological state. For linear chains, or even comb-like ones, the 'topology' is not fixed in the mathematical sense of knots. But if polymers are made with their ends joined into rings, as is now chemically feasible, then their mutual topology is fixed for all time. Two rings are either linked in a number of possible ways, for example, or unlinked (see Figure 17).

(a)   (b)

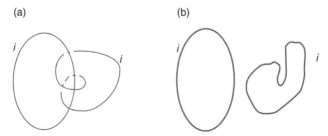

17. The two-ring polymer chains, i and j, are permanently (topologically) linked in state (a) (with in this case a 'winding number' of 2), and unlinked in state (b).

The 'ring melt' has as a result many extraordinary properties. If its constituents are synthesized as rings before being concentrated together, so that they are permanently unlinked, then the polymers form a fluid, each chain able to diffuse around its neighbours. If, however, they are linked in an overlapping state, then they form a structure like a three-dimensional molecular chain mail. Although there are no rubber-like, or vulcanized, cross-links, this material is a solid—it cannot flow. To be able to tune between liquid and solid with the aid of no more than long-range topological quantities, such as linking, is a remarkable property of soft matter.

Chapter 4
Gelification and soapiness

The day that I was visited by two puzzled chemists became a memorable one. The professor and his PhD student had been interested in synthesizing a small molecule arising in biological processes—a 'peptide'. The word was unfamiliar to me as a physicist and I needed a simple explanation. I got one: peptides are very short versions of biological polymers called proteins. They are, like those larger macromolecules of life's structure, built of linear strings of 'monomers', small molecular building blocks analogous to those of synthetic polymers. In the case of proteins, however, the monomers are not typically identical but chosen from a limited alphabet of twenty 'amino acids' within the same protein. My colleagues had been attempting to make dilute solutions of a particular peptide—a short string of twenty-seven of these building blocks—but had noticed something peculiar when the synthetic procedure had made some errors in the choice of some of the amino acid building blocks. Rather than forming the expected liquid solution, the solvent had simply 'gelled' into a soft transparent solid. Although 98 per cent fluid, the test tube could be turned upside down yet cause nothing to flow out.

Thinking through the problem identified the core issue of *connectivity*: a material can behave as a solid providing every part of it is connected to every other by some rigid (or at least elastic) route. In hard solids such as metals and glasses there are many such routes—the entire volume is filled with rigid bonds. But soft gels exploit the possibility to use only a small fraction of their volume to impart rigidity, providing that the 'rigid' elements are of lower dimension than the solid. Foams are rigid because they connect three-dimensional space with two-dimensional cell walls; gels because they connect through space with (one-dimensional) polymers. The 'random rubber' of connected polymer chains in Chapter 3 showed how polymer solutions, even at very low concentration, may exhibit elastic properties that respond principally to the presence of the long polymer chains, rather than to the flow properties of the solvent that surrounds and supports the polymers.

Peptides, however, are neither the extended sheets of foams nor the long polymers of gels. They can be thought of more like short flexible rods, so do not explain on their own the solid nature of my colleagues' solution that refused to flow. But if they had somehow spontaneously arranged themselves into long chains, that would explain the appearance of the jellied state. The challenge was to understand how extended, low-dimensional objects like strings and sheets might form spontaneously from individual molecular units. This brings us to the world of 'self-assembly', a third class of soft matter. Like the colloids of inks and clays, and the polymers of plastics and rubbers, 'self-assembled' soft matter also emerges as a surprising consequence of Brownian motion combined with weak intermolecular forces. Like them, it also leads to explanations of a very rich world of materials and phenomena, such as gels, foams, soaps, and ultimately to many of the structures of biological life.

One-dimensional self-assembly

The example of the gel-making peptides turns out to provide an object lesson in the self-assembly of extended mesoscopic tapes. The short peptide rods of amino acids do not react together chemically, as do the monomers of the permanent polymers of styrene or ethylene of Chapter 3. But there are other, weaker attractions that may endow potential monomer units with the mutual 'stickiness' that causes them to link up. The most important in this example, and in many other self-assembled polymers, is the 'hydrogen bond'. Illustrated in Figure 18, it arises between nearby hydrogen and oxygen atoms through the sharing of an electron between them. Significantly, every amino acid potentially offers both a hydrogen and an oxygen atom to its near-environment, and an extended peptide molecule presents possible hydrogen bonds on both of its sides. This means that, like dancers in a reel holding out both hands as they move to the music, it is possible to make and break long chains that weave and dance around each other. The short rods of the peptides join up so that rod lies against rod—their length forming the width of an extended tape.

In 'polar' solvents—that is, solvents whose molecules contain electric charges, like water—the strength of each bond is a little larger than the average thermal energy $k_B T$ at room temperature that we met in Chapter 1. Since this is, therefore, not much greater than the energy in the thermal kicks that they receive from their environment, the links that they forge will not be permanent, as with the much stronger bonds of permanent polymers, but will occasionally be broken by random thermal motion. Figure 18 shows that the links between adjacent peptides can be built from several hydrogen bonds, however. The event of all the hydrogen bonds connecting two peptides simultaneously breaking is much rarer than the event of a single bond break. This means that their self-assembled tapes can grow very long before a break occurs at some point along their length.

Such random but reliable probability of broken bonds, their continuous making and breaking due to Brownian motion, and the identical chemical nature of each of the self-assembling units, together determines the likelihood of finding any possible length of string in the solution. This probability turns out to be an exponentially decreasing function of the length (or equivalently of the number of units). Just as radioactive substances have a 'half-life'—the interval of time after which just one-half of the radioactive atoms present at its beginning persist undecayed at the end, so self-assembled polymers possess a 'half-length'. Within every extra length of this amount, the probability that the polymer continues to survive unbroken is exactly halved. The 'half-length' of the self-assembled polymer is also a good measure of the average length of the entire population. The emergence of such well-defined statistical descriptions of self-assembled structures in soft matter is the first of two ways in which Brownian motion constitutes the key to self-assembly.

Brownian motion acts within self-assembled soft matter in a second vital way—providing the very background dynamics that generates self-assembly in the first place. Weak attractions, such as hydrogen bonds, are typically very short-ranged. Two peptides have to be in close proximity before the bonds will form in the first place—like the dancers in the reel, they have to be within two arms' length before there is even a chance of holding hands. At first sight, this leads to an insuperable problem, for in dilute solution the average distance between them is much greater than the distance at which they begin to feel each other. There seems to be a need for an invisible agent at work, bringing the units close enough together for them to stick, building the long strings in sequence. The 'agent' is none other than the Brownian thermal motion once more, for in the microscopic world every object is in constant thermal agitation. Not only will individual peptide units be translating randomly, exploring space through diffusion, but also their orientation will tumble, turn, and twist in a 'random walk of direction'. This orientational diffusion is important: in Figure 18(a) the new addition to the growing string of peptides must not only present itself at its broken end but with the correct orientation as well (lying parallel to the last peptide of the assembly).

If it were possible to watch a microscopic movie of the fluid containing the self-assembling peptides, there would be constant,

18. (a) The addition of a peptide onto the end of an existing stack satisfies several hydrogen bonds between exposed hydrogen and oxygen atoms on neighbouring peptides, allowing chains of peptides to grow.

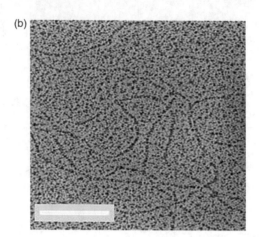

(b)

18. (b) A cryo-electron micrograph of the assembled 'peptide tapes'. The scale bar shows 100 nanometres.

frenetic activity. The extended tapes would constantly be breaking and reforming at all places along their length, while they twisted and turned in the thermally active solvent. Peptides would add and peel off from the polymer ends, diffusing randomly between the many chains, tumbling and twisting so that they sometimes added on to a chain end when they encountered it, but more often making no permanent bond and dancing away again. Yet, in spite of all this complexity, there are clean and simple predictions we can make about how many peptides will be accommodated into chains of any length.

Peptide and protein fibrils

The one-dimensional self-assembly of the short peptide molecules has developed into a fruitful soft matter story. Although their simplest form of concatenation can be understood by the simple, equilibrium model of linear addition, there is a special feature of the giant assemblies generated by the rod-like form of the peptide units: the polymers are more properly described, as we have seen, as long 'tapes' rather than 'strings'. Joining rods together side-by-side creates a flexible structure that looks more like a metallic watch-strap, rather than a string of pearls. Alternatively we might say that the resulting polymers have *two* measures of width rather than one—a long diameter given by the length of the rod-like peptide units and a short one given by the size of a single amino acid link.

From this simple additional observation, and two further properties of the peptide tapes, emerges a rather elegant set of geometrical properties of their self-assembled state. First, the structure of amino acids mean that the chemical groups presented each side of the tapes are, in general, different. Making the sides of the tapes different in this way means that, without any specific knowledge of the forces and shapes of the molecules, the tapes will tend to possess an average net curvature towards one side and away from the other. Like a watch-strap around your wrist, it will adopt a natural curvature.

The second salient property is that amino acids are not equivalent to their mirror images (see Figure 19(a)).

Such molecules are said to be 'chiral' (from the Ancient Greek *chier* for 'hand'). Like left and right hands, amino acids are not equivalent to their mirror images. Remarkably, the amino acids produced in living systems are all of the left-handed 'L'-form, none of the right-handed 'D-form', of Figure 19(a). Such bias to 'handedness' at the molecular level typically induces a similar handedness in the emergent structures they form collectively (we will encounter a further example of handedness from the field of liquid crystals in Chapter 5). In the case of filaments or tapes, chirality emerges as a twisting of the lateral structures along their length—like a screw-thread they may twist in a left-handed or right-handed sense, but which will be determined by the handedness of their smallest building blocks.

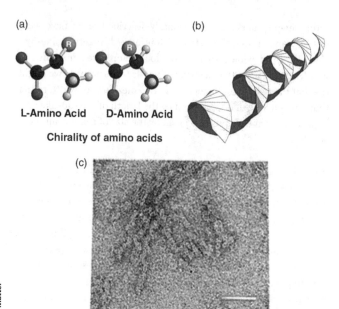

(a)

L-Amino Acid   D-Amino Acid

Chirality of amino acids

(b)

(c)

19. (a) The structure of amino acids: dark filled circles indicate carbon atoms, grey nitrogen and open circles oxygen; the residue marked 'R' represents the variable side group that is the only component that changes between different amino acids; the L- and D-forms are mirror images of each other; (b) diagram of a self-assembled peptide tape with both bend and twist—the distinct sides are shaded light and dark; (c) an electron micrograph of self-assembled peptides with chirality and bend.

Now imagine combining a gentle curvature of the self-assembled tape with a simultaneous twist. The result is in general the spiral 'tress' of Figure 19(b) as the tape twists as it curves, reminiscent, perhaps, of an uncomfortable roller-coaster track. The structures visible in the electron micrograph of Figure 19(c) are of precisely this kind. They are composed of thousands of rather complex molecules, the peptides, each built from hundreds of atoms. Yet with a few simple ideas such as linear self-assembly, curvature, and handedness, it is possible to understand how such elegant structures emerge from the random Brownian soup of their constituents.

Understanding such hierarchical self-assembly of peptides has an important medical significance. There are a number of diseases that present self-assembled aggregates of proteins, or clipped fragments of proteins, in the tissues affected. Sometimes these take the form of fibres and fibrils, others in larger and less linear 'amyloid' plaques and other aggregates. Alzheimer's and Huntingdon's disease are examples of pathologies in which the structures of Figure 18(a) form the core. Several laboratories around the world are engaged in searching for ways to disrupt the pathways to self-assembly.

At other times, the local self-assembly of proteins is a convenience. The expression used by the chemists, who had asked for advice on their 'gelled' peptide solution, recalls another everyday miracle of solidification in the making of table jelly. The active ingredient is the animal protein gelatin—an example of a bio-polymer that at the relatively high temperature of freshly boiled water not only dissolves but opens up into the extended random chains we examined in Chapter 3. As the solution cools, clusters of potential weak hydrogen bonds—the same type that linked the peptide tapes together—begin to work towards assembly in this case also.

But gelatin is a long-extended polymer, not a short peptide, and the groups that are able to form hydrogen bonds are clustered in relatively short sections of the chain. This predisposes the proteins to form short local assemblies of three separate gelatin chains, joined in the form of short triple helices (see Figure 20). Flexible gelatin chains link these helical regions together, so bridging across the space of the solution and creating the cross-links responsible for the emergent elasticity of the solid jelly. Even though the water occupies the great majority of the volume threaded-through by the gel, the liquid nature of the solvent cannot prevent the emergence of solidity across the self-assembled structure.

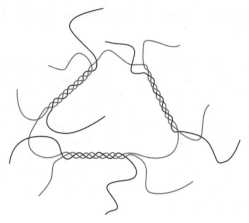

20. A schematic structure of a gelatin network after dissolution of the protein polymers then cooling. Hydrogen bonds cause the local formation of (helical) junctions within the network.

From one dimension to two: soapiness, surfactants, and membranes

The sense of 'soapiness' opens up a world of familiar, but nonetheless fascinating, phenomena. The ability to clean dirt from skin and clothes with this remarkable material, its slippery smoothness in solid form when wet, the formation of soap films, bubbles, and foams, and even, as we shall see, the sluggish viscosity of shampoos are all further examples of self-assembly in soft matter. The central difference between the peptides, proteins, and gelatin we have looked at so far, and the underlying self-assembled structures of the soapy substances, is that the first are one-dimensional and string-like, while the second are two-dimensional and sheet-like. Surfaces may self-assemble, given the right building blocks, as well as lines.

Soap is also another example of ancient soft matter technology. The oldest record of a recognizable process for soap-making is recorded on a clay tablet inscribed in ancient Babylonia around

2800 BC. The ancient Greeks and Romans knew about soap-making, as did the ancient Chinese. There are records from the medieval Arab Middle East of recipes for soaps as early as the 9th century, and a 12th-century document describing the use of ashes (*al-qaly*) has given us the root for the modern English word *alkali*. The classic process of soap-making typically extracted potassium hydroxide (a very strong alkali) from the ashes of burnt plants. The alkaline substance was then heated with fatty acids extracted from animal or vegetable oil. The reaction produces the molecular units of soap—the so-called 'metal salts of the fatty acids'. They have tadpole-like structures such as that shown in Figure 21. The 'head' of the molecule is salty—a charged and bound metal atom (an 'ion'); the tail uncharged and molecularly identical to a fat or an oil.

This generic structure of soap-forming molecules codes for a natural self-assembly in solution. The two ingredients of the soap-making process endow two very different properties to the molecule. From the fat comes the 'tail' of carbon atoms with side-bonded hydrogen pairs, shown in Figure 21(a). This is the ubiquitous structure of oils, greases, and fuels. Fats do not, of course, dissolve in water—they belong to the 'hydrophobic' ('water-hating') class of molecules that resists contact with it. On the other hand, the fatty tails are attracted to, and will dissolve in, liquids that have an oily structure themselves. The metal alkali ash, on the other hand, is the provenance of the charged group that forms the 'head' of the molecule. This is hydrophilic or 'water-loving' due to the polarizability of the water molecule when close to an electric charge. Figure 21(b) coarse-grains the representation of the soap molecule—clarifying its two-natured, or 'amphiphilic', character. Figure 21(c) indicates the form of self-assembly that is responsible for the cleaning action of soap: a small oil droplet binds weakly to the oily tails of the amphiphiles, creating a structure that also removes those hydrophobic tails from contact with water solvent, while their charged heads are the only part of the cluster exposed to water. It also explains the

(a)

(b)

Ionic part
(–COO⁻ Na⁺) $\longleftarrow$ $\longrightarrow$ Hydrocarbon part

(c)

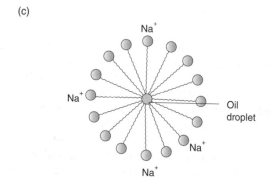

21. (a) Atomic representation of a sodium salt soap (surfactant) molecule; (b) a coarse-grained representation; (c) their self-assembly around an oil droplet.

standard name given to these soap-forming molecules—
'surfactants': they are active at surfaces between the liquids of the
oily and polar characteristics of their own two natures.

The classic soft matter motif of coarse-grained structure has
appeared once again. The cluster of surfactant molecules around
the oil droplet possesses length scales of 1–10 nanometres—large
enough to acquire properties that depend only slightly on the

exact atomic and chemical structure of its components but small enough to be in thermal, Brownian equilibrium with its environment. Such structures, called 'micelles', often approximately spherical in form, have most of the properties of the colloidal particles of Chapter 2: they form 'colloidal atmospheres', may undergo phase transitions into colloidal liquid and crystal forms (if sufficiently narrowly distributed in size), and if sufficiently large will scatter light, forming the cloudy liquids familiar towards the end of washing up dirty dishes.

The structure of Figure 21(c) is suggestive of a type of self-assembly that might take place even without the presence of oily droplets to clean away. Providing that the surfactant molecule tails are sufficiently hydrophobic, the micelle structure will be effective in maximizing the exposure of hydrophilic components (molecular heads) to the water and in shielding the hydrophobic parts (molecular tails), even without the oil or fat droplet in the centre. The most important criterion that has to be satisfied to make this work comes from the geometry of packing. The surfactant head groups need to cover the surface of a sphere of micelle radius r (area $4\pi r^2$) whose volume ($4\pi r^3/3$) is occupied by their tails. So if the ratio of surface to volume ($3/r$, when you do the division) matches the ratio of 'head' and 'tail' volumes in the molecule, for some size of sphere r then it is likely that that size will predominate in a spontaneous process of self-assembly into such nanoscopic micelles. Other ratios of head and tail volumes will favour the emergence of correspondingly different assembled structures with less curvature, such as rods or sheets.

The surface-to-volume criterion for micelle formation is not quite the only criterion, for if the head group is too small there is a penalty against the strong curvature of a micelle that compresses the tail groups and provides more room for the heads. In this case the drive to bury the hydrophobic tails out of contact with water can be satisfied by other structures. The most significant is a double layer, or 'bilayer', of amphiphiles, for then the

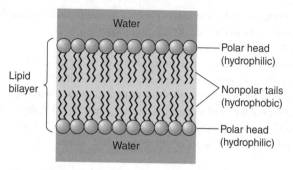

Water

Polar head
(hydrophilic)

Lipid
bilayer

Nonpolar tails
(hydrophobic)

Polar head
(hydrophilic)

Water

22. Schematic illustration of an element of surfactant bilayer membrane, in this case whose lipids possess two hydrophobic tails. Individual water molecules are not shown, only the region external to the bilayer where they are found.

self-assembled structure takes the form of membranes or sheets built from these extended bilayers (see Figure 22). Such two-dimensional structures may extend laterally to length scales many orders of magnitude higher than their widths.

Membranes are the two-dimensional equivalent of the one-dimensional polymers of Chapter 3. They may, like the polymers, endow a liquid solution with a soft rigidity, by virtue of making connections across the space of the solution. When these connections are polymers, such soft solids are called gels; in the case of internal membranes they form the structured materials we know as foams, or at very small scales, 'microemulsions'. The self-assembling molecules construct the sheet-like faces of the cells in foams, rather than their string-like edges in the case of polymer gels.

The membrane structure of Figure 22 possesses a 'front-back' symmetry—it looks the same when flipped upside down. For this reason there will be no net tendency for the sheet to curve in either direction. However, just as in the random coiling of polymers, rather than the rod-like rectitude that Staudinger expected of them, this does not mean that the sheets will be perfectly flat on all length scales—it constrains only the very local geometry to be flat. Membranes, like polymers, will also exhibit curvature away from their mean orientation at any one point, in consequence of the thermal environment. Curved states of the membrane will require some energy to access, but, as for polymer solutions, this is available from the solution through Brownian motion, and as usual the typical thermal energy acquired by any single mode of deformation is the thermal energy unit $k_B T$. As a result, membranes in solution also wander away from the flat smooth shape they would adopt at a temperature of absolute zero.

Thermal fluctuations in membrane curvature and shape generate an intriguing form of effective repulsion between pairs of membranes. In another guise, they also give rise to an elastic-like resistance of single membranes to being squeezed flat. Just as with rubber elasticity, this emergent force is entirely thermal in origin—there are no electrostatic, electronic, or chemical contributions. Akin to the rubber-elastic force from stretched polymers, it arises instead from the restriction of possible configurations of the membrane (in technical language, a reduction in their *entropy*). In the case of membranes, confinement forbids many configurations that a free membrane would otherwise have explored (see Figure 23).

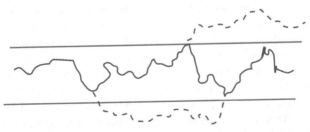

23. **Schematic depiction (side-on) of a fluctuating membrane confined between two walls. Disallowed configurations are shown dashed.**

The smaller the gap, the more frequent are the points at which the ruffled membrane encounters the surface. As each frustrated attempt to cross the surface costs the membrane a single configurational constraint, it contributes a consequent effective energy penalty of just the familiar thermal quantity $k_B T$. All of these encounters between the membrane and the wall that confines it add up to a thermal force with which the membrane pushes outwards. A familiar consequence of this repulsive force, named after Berlin physicist Wolfgang Helfrich, who first proposed and calculated it, is the splitting and cracking of soap when it is wet. Locally, surfactant molecules pack into stacks of bilayers in cakes of soap, however random the structure is on macroscopic scales. Exposing the solid soap to water opens up the possibility of the bilayers exploring multiple fluctuating configurations, but only if water is taken up into the interstices between them. The repulsive force pushing the bilayers apart draws more and more water in and allows, eventually, macroscopic cracks to grow in the soap. At a much lower exposure to water, its incorporation between the bilayers leads to the slippery feel of wet soap. Membranes may slide over each other easily in their transverse, in-plane, direction, when lubricated with the water drawn between them by the entropy of their out of plane fluctuations.

All real membranes are, of course, finite, so one might think that they must always possess edges. However, bilayer edges floating in water must come with a high penalty, for their hydrophobic tails are still exposed to the solvent there. In practice this state can be avoided by the adoption of a small local curvature in the bilayers—for then they may remove all edges by adopting the finite but boundary-free surface of a sphere. Microscopic droplets enclosed by bilayers in this way are called vesicles (see Figure 24). By virtue of their bilayer, they possess internal and external spaces, both of which are aqueous but disconnected. Vesicles may therefore serve as containers for solutions that differ in their content from the external solvent. Providing that they are large enough the small

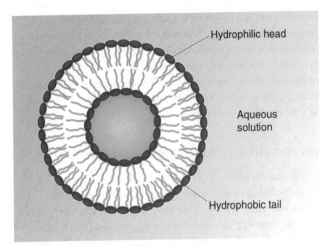

24. Schematic depiction of a vesicle composed of a surfactant bilayer.

curvature required to form them can be negligible compared to the available thermal energy.

These structures are highly significant in biology, for they are the chosen base structure of cells. In animals and single-cell organisms, cell walls are lipid bilayers. The self-assembled nature of these essential membranes is literally vital in the context of life as no special machinery is required to construct the outer wall of biological cells. Once intact, cell membranes permit different concentrations of salts such as calcium and potassium to occur between their interiors and exteriors. These concentration differences may then drive activity on the cell surfaces and interiors. Biology also recruits another special, if counter-intuitive, feature of surfactant bilayers. For although they are resistant to tearing apart and bending, so in that way they resemble elastic sheets, yet within the two-dimensional space of the membranes themselves they are fluid. A body embedded within a bilayer may diffuse freely, the amphiphilic molecules flowing around it like a two-dimensional liquid. Many of the signalling and biochemical processes essential to life are actually performed by protein complexes, themselves self-assembled within the surface of cell membranes by such two-dimensional diffusion, followed by aggregation. Sometimes these assemblies act back on the membrane, curving and deforming it even to the extreme case of 'budding', when sections of membrane curl up into vesicles and leave the membrane altogether.

Two curvatures, exotic phases, and topology

If a self-assembled bilayer is identical on both surfaces, so symmetric about its mid-plane, it has, as we have seen, no tendency to curve toward either side. When that symmetry does not apply, however, the resulting tendency to bend can lead to other emergent soft structures. Some of these consist of separate units, such as the micelles, but others embed continuous sheets in some quite exotic structures that we explore in this section.

An important aspect of curvature in two dimensions rather than one, that we have not yet considered, is the presence of two independent 'curvatures' at every point. On surfaces there are two directions along a sheet, at right angles to each other, along which curvature may defined. These two curvatures may have not only different sizes but also different signs. The three distinct choices of combination of signs differentiate between three qualitatively distinct possible structures: a local minimum, or valley (when both are positive); a local maximum, or peak (when both are negative); and a saddle (when there is one curvature of either sign), all visualized in Figure 25.

The magnitude of the curvatures is conveniently described by the radii of curvature R_1 and R_2—the radii of the circles whose curvature matches that of the membrane locally. The more curved the surfaces are, the smaller is the relevant radius of curvature (so the curvature in direction 1 will actually be $1/R_1$). It is always possible to define the two curvatures in this way, but there is a

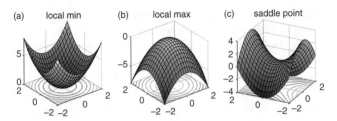

25. The three possible sign configurations of local curvature on a membrane: (a) minimum; (b) maximum; (c) saddle.

more helpful physical way to group them, through constructing their means. There are two ways of constructing a mean, or average, of two quantities—the arithmetic and geometric mean (remember the arithmetic mean of two numbers a and b is $\frac{1}{2}(a+b)$, while the geometric mean is $\sqrt{(ab)}$). It turns out that this high-school mathematical fact has a remarkable interpretation within the physics of curved membranes. The 'mean' (arithmetic) curvature κ_1 depends on the sum of the two curvatures—so $\kappa_1 = \frac{1}{2}(1/R_1 + 1/R_2)$, while the so-called 'Gaussian' (geometric mean) curvature κ_2 is defined through their product via $\kappa_2 = (1/R_1 R_2)$. We might note that the Gaussian curvature has a special property in reference to the three local geometries of Figure 25. While positive for both minimum and maximum, it is only negative in the case of a saddle.

The Gaussian and mean curvatures respond, therefore to different geometries of membrane curvature—saying that a membrane forming molecule tends to form saddles is equivalent to a statement that it favours low (negative) Gaussian curvature.

We have identified the relevant, universal, and coarse-grained variables, characteristic of soft matter, for the case of membranes. It's all about the curvature. There is a further wonderful and beautiful mathematical property of the Gaussian curvature that connects its apparently local definition to a global property of the object that the membrane covers. Known as the Gauss–Bonnet

Theorem, the result is easy to state (though not so simple to prove): if all the local Gaussian curvatures of an entire membrane of a set of self-assembled objects are added up over the area of the membrane, the result (when divided by 4π) is simply the number of disjoint objects minus the total number of holes in them. To clarify by example, a spherical micelle is a single object with no holes so the Gauss–Bonnet sum just gives 1, while a doughnut-shaped surface is also a single object but contains one hole: the Gauss–Bonnet sum give 1–1 = 0.

The number of such 'holes' in a body is a 'topological' property—like the entanglement constraints of the polymers in Chapter 3, it does not depend on shape, deformation, or size but can only be changed by cutting and/or joining. A doughnut and a coffee cup share the same topology (thanks to the handle) but possess very different geometries. The Gauss–Bonnet sums over their surfaces will be identical and both equal to zero.

It is possible to generate conditions that favour continuous, curved, and highly complex topologies in surfactant-oil-water systems, by choosing molecular structures for the surfactants that enhance the Gaussian curvature of their assembled membranes. The structures that result are multiply connected—consider, for example, the surface describing six tubes joined smoothly and at right angles at a vertex. The curvature around the cylinders is always in the opposite direction to the curvature that joins one cylinder to its neighbour—the surfaces around the junctions are locally saddle-shaped. The Gaussian curvature is negative everywhere, so, for example, a molecular structure of amphiphile that favours negative curvature of the membranes it builds will tend to self-assemble into structures that contain high concentrations of such vertices.

One might on first reflection assume that all such phases would maintain the high entropy of a random distribution of surfactant tubes and meeting points. Such disordered, or 'sponge phase', assemblies do indeed exist, but, in a similar surprise to the existence of hard-sphere colloidal crystals, crystalline phases of bilayers with negative Gaussian curvature also form spontaneously. An example is shown in Figure 25, together with their macroscopic ordered crystal forms observed by light microscopy.

Colloquially known as 'plumbers' nightmare' structures, for obvious reasons, the continuous and regular self-assemblies have been found in many different crystal symmetries. The single-surfactant, lipid molecule monoolein, for example, self-assembles into a repeated three-dimensional 'climbing-frame' stack of tubes and their junctions, called a 'cubic phase' (depicted in Figure 26). Changing external conditions can exert delicate control over the types of infinite structures that self-assemble. Crystals with a different, tetrahedral, symmetry appear at higher concentrations of surfactant. At higher temperatures, on the other hand, hexagonally packed arrays of extended rod-like micelles are preferred to the crystals, as the stronger fluctuations in the lipid tails erodes the stability of the negative Gaussian curvature.

(a) (b)

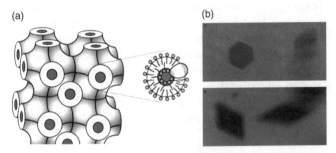

26. **Amphiphilic cubic phase used to embed membrane proteins in a regular array** (a) schematic of local cubic structure, with an insert displaying structure at the next smallest scale of lipid molecules (and the possible configuration of a protein molecule inserted into the membrane); (b) micrographs of self-assembled crystals of the lipid monoolein.

Self-assembly of amphiphilic polymers

A strong theme throughout this chapter has been the link across length scales between the molecular structure of self-assembling molecules and the emergent forms of their assemblies. We might say that their local structure *encodes*, in some sense, the forms of their self-assembly. The simple head-tail nature of surfactant molecules, for example, implies the formation of separate spatial regions dominated by either head-head or tail-tail contacts, supplied by the emergence of bilayers.

The local formation of bilayers is a 'first-order' requirement of self-assembly. The bilayer membranes in turn, once formed, can take on many geometrical forms, depending on the 'second-order' molecular determinants such as preferred curvature and surface-to-volume ratio. The examples of soft matter self-assembly presented so far have been based on relatively small molecular building blocks: peptides and surfactant lipids. But the same game can be played using much larger elements. Polymers themselves can be manufactured to display an amphiphilic character analogous to that of surfactant molecules. It is possible, for example, to synthesize polymers whose chains are composed of two sections, each concatenating a different monomer called a 'block co-polymer' (see Figure 27). Polyisoprene-Polystyrene is one common example, in which the monomers possess rather different chemical character. This in turn predisposes an attractive interaction between chemically similar monomers and a net repulsion between different kinds: A-monomers prefer to surround themselves with other A-monomers, and likewise for B-monomers.

27. Schematic diagram of an amphiphilic block co-polymer composed of a string of A-monomers (filled circles) bonded to a string of B-monomers (open circles).

Melts of A-B-co-polymers are therefore candidates for encoding self-assembled structures in the same way that surfactants do but at much larger length scales. Furthermore, since the self-assembling components (polymers) are now already very large compared to atoms, the basic characteristics of mesoscopic soft matter imply that a degree of universality might be expected in the emergence of structure. At the coarse-grained level of polymer physics, arranging the macromolecule across an interface between A-rich and B-rich regions of the polymeric fluid so that each can coexist in spatial regions rich in their own chemical kind, might win on energetic grounds. However, in stretching out so that each half occupies a different spatial region, the polymers would have to pay the same entropy cost that stretched chains in a rubber do. Such delicate balances of entropic disorder opposing an energetic ordering tend to lead to interesting structures. A few variables that will control the form of emergent structure also suggest themselves: for the simple A-B-co-polymer they are the two-parameter set of the ratio of A to B in the chains and the temperature (which controls the effective A-B interaction).

The self-assembly of even this simple diblock co-polymer system exhibits a remarkably rich world of possible forms in terms of these variables.

Figure 28 contains cartoons of the principal phases presented by melts of diblock co-polymers and maps of where they appear in terms of A/B fraction and the strength of A-B repulsion. As one might expect, when the two components do not repel strongly, the entropy of random chains wins and the fluid is well-mixed and disordered (structureless). But as A-B contacts are made less and less favourable, a set of locally demixed structures appears, whose form most strongly depends on the A/B fraction in the chain. When A and B both occupy about half of the chains, symmetrically, then a symmetric nanostructure assembles: layers or 'lamellae' of alternating A and B that constitute polymeric equivalents of stacked lipid bilayers. But as the polymers become

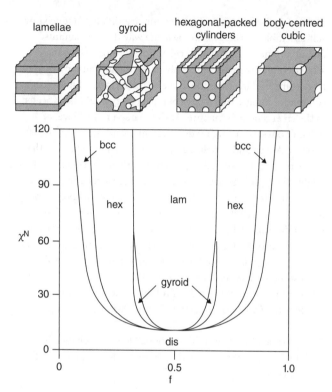

28. **Schematic map of self-assembled structures in diblock A-B co-polymers in terms of the fraction of A-monomer in the chain (x-axis) and the effective repulsive interaction between A and B (y-axis); (c) schematic forms of the crystal (bcc), cylindrical (hex), gyroid, and lamellar (lam) phases that appear.**

less symmetric, so do the self-assembled structures, and the A-B interfaces start to curve more towards the minor component. As A-monomers reduce below about 30 per cent of the chain, they adopt cylindrical regions, arranged hexagonally in the planes at right angles to the cylinders—two-dimensional crystals. In a small region of the control variables, an exotic structure appears of interwoven tube-like regions of A and B in a lattice called the 'gyroid'. At even lower A-fractions a fully three-dimensional crystal of solid order is preferred. These crystals, like the cubic phases of surfactants, are bizarre to contemplate in that their local molecular dynamics are entirely liquid—the solid order emerges only at larger length scales, supported by the constant Brownian motion of the liquid components at length scales far below.

Flow of self-assembled soft matter—and sluggish shampoo

If self-assembled soft matter exemplifies the general features of universality and meso-scale structure we identified in Chapter 1, it also illustrates its characteristic dynamical behaviour. Large soft structures at the nano-scale possess slow dynamics, and the emergent forms of self-assembly are no exception. They have some special features worth exploring briefly, not found in the fixed forms of polymers and colloids.

Self-assembly in one-dimension leads, as we have seen, to an ensemble of polymers. Whether these are locally tape-like or string-like, at larger length scales their dynamical properties would be expected to lie close to those of permanent macromolecules. We have already seen in Chapter 3 that single-chain dynamics are controlled chiefly by connectivity and many-chain fluids dominated by uncrossability—the domain of entanglements and 'reptation' in effective topological tubes. This remains true for self-assembled polymers, but with one important difference. Self-assembly comes with its own dynamical structures: it is an equilibrium phenomenon, underpinned by continual motion, change of state and fluctuation. For polymers, the special length-distribution emerges from continual breaking and recombination of chains.

The internal dynamics of one-dimensional self-assembly—a system that has been termed 'living polymers'—does make a difference, especially to the motion of chains in entangled states.

This is because chain ends have a special role to play in disentangling old configurations and exploring new ones, the route by which an entangled fluid returns to equilibrium after flow. As we saw in Chapter 3, it is only chain ends that have complete freedom from the entanglement field around them—they create new 'tube' constraints by diffusion and release old ones. The difference now is that self-assembled polymers may create new free ends anywhere along their length by simply breaking.

Provided that the breaking and recombination are considerably faster than the total reptation time of a chain, the process of relaxation of elastic stress now typically proceeds as shown in Figure 29. It is not necessary, as with fixed polymers, to wait for a whole reptation time for a current chain end to annul the entanglements around a segment of tube. For now a break is likely to occur in the vicinity long before then, offering an opportunity to disentangle after a much shorter diffusion path for the chain. This process was proposed and confirmed in a beautiful 1990 paper by British theoretician Mike Cates and French experimentalist Jean Candeau. A stark and surprising prediction was that the reptation-reaction process made stress relaxation 'democratic' among all chain segments. The nearest distance of all segments to a chain end would typically change multiple times in the process of a single relaxation time. As a result, stress held by chain segments closer to chain ends would no longer relax faster than those buried towards the centre. The consequence for rheological measurements of stress-relaxation is that there would only appear a single relaxation rate.

Experiments confirmed this almost 'model' behaviour very accurately in a system of simple surfactants in water with added salt. There is a preferred curvature to the self-assembled monolayers which prefers an emergent morphology, not of spherical but of long *cylindrical* micelles. When the conditions are right, these may extend to very great lengths but also adopt fast

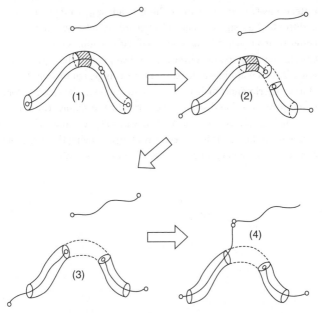

29. 'Living reptation' in shampoo containing worm-like surfactant micelles: the stress held by an entangled tube segment (1) (shaded) is typically released by a chain end formed in a nearby breaking event (2), followed by reptation through a short distance before a recombination with a new chain in its neighbourhood (4).

breaking and reformation kinetics between their end-caps. They are known as 'worm-like micelles', since at long length scales their gentle thermal bending will result in random walks threading the solution.

The technological application of this subtly balanced system of structure and dynamics can be found in most bathrooms. Shampoo is, like all soaps and detergents, a solution of surfactant molecules in water. That is all that is necessary for its activity, but marketing surveys have for a long time told its producers that customers prefer a highly viscous, even viscoelastic, product to a watery one. One way to achieve this is to add a neutral polymer additive to the soap as a rheological modifier, but how much better to arrange it that the soap itself self-assembles into entangled polymers, so delivering the cleaning and the high viscosity (even at high-water content) in a single component? Home experiments (such as cooling a shampoo in the refrigerator to temperatures that disfavour the worm-like assembly) can verify that the high viscosity can be made to vanish by a change of external conditions. It is a direct way of experiencing the delicately controlled and emergent properties of soft self-assembly.

Chapter 5
Pearliness

This afternoon was very cold and when I returned home the heating wasn't yet on. So, the first job was to punch the 'Advance' button on the heating control unit by the boiler. The display lit up immediately, informing me in clear dark letters and numerals the time now, the time the heating would stay on until, and the current temperature. The figures updated before my eyes.

We take these once-astonishing electronic displays for granted now, but the idea of text that could be switched on electronically as a living page, continually rewritten, would once have seemed the stuff of fantasy. What marvellous material is it that can support temporary text in this way? The answer lies in our next class of soft matter—the liquid crystals.

Rod-like structures and orientational order

The classification of soft matter, which takes its cue from the shape of the underlying mesoscopic component, suggests a class of soft materials that would possess properties distinct from those of both colloids and polymers. For between the compact near-spherical particles of the first and the flexible strings of the second ought to lie molecules that are extended (like polymers) yet rigid (like colloids). Nanoscopic or molecular *rods* provide the theoretical notion that animates the exploration of the soft matter class of 'liquid crystals' and motivates the first step into thinking about this next class of materials. Immediately evident is a new variable to think about at the local level, in addition to the density and possible crystalline spatial ordering that the colloids required. For in this new case the fundamental structures possess a special direction as well as a position—their long axis points somewhere. Such an orientational variable means the substance can adopt ordered or disordered states, depending on how correlated the molecular rods are with each other. Figure 30 considers two imagined cases of local regions in a fluid of rod-like entities, disordered and ordered in such an orientational sense.

Were a local directional order to emerge, it would have to select a preferred direction, which has been termed the 'director'. A single rod in a region where there exists a dominant orientation explores this disposition of its neighbours by diffusing among them, and so becomes constrained to adopt their preferred direction as well—it is 'directed' by the directionality of rods in its vicinity.

Such a liquid with inherent directionality might display many distinctive properties. One type we can anticipate right away, for

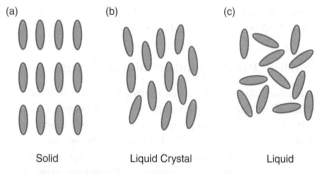

(a)   (b)   (c)

Solid   Liquid Crystal   Liquid

30. Rod-like molecules in thermal agitation might display order (centre) or disorder (right) in their orientation, irrespective of their spatial (crystalline) order (left) in the solid state.

light also has a direction. As James Clark Maxwell first realized in 1865, light is a self-sustaining wave of orthogonal electric and magnetic fields. Its 'polarization' indicates the direction perpendicular to the propagation of a light wave in which its electric field points. Polaroid sunglasses work because their material possesses a special direction, allowing light of (usually) vertical, but not horizontal, polarization to pass through. Since light interacts with the material it passes through like this, the orientational order within a fluid composed of rod-like particles would be expected to display unusual optical properties, especially when the light used to examine them is polarized. We would, in particular, expect liquid crystals to display special optical properties.

A strange fluid and a double melting

Equipped with these useful conceptual ideas—gifts of hindsight naturally—of a fluid with a local pointing direction and its likely ability to interact with polarized light, we may now accompany Austrian chemist Friedrich Reinitzer to his laboratory at the Institute of Plant Physiology in the Prague of 1888. Reinitzer was running a series of experiments on the biological molecule cholesterol (beloved of cardiac health conversations), with the goal of determining its molecular weight, at that time still unknown. Having extracted and purified the species in a concentrated solution, he first set about examining its melting behaviour. The experiment slowly increased the temperature of the sample, looking out for the extra flow of heat into it during periods when the temperature rise pauses and which signifies a melting point, and observing the physical form of the material throughout.

The low-temperature solid form of cholesterol did indeed melt, at 145.5 °C, but into a strange liquid, of a pearly milkiness in appearance rather than transparent. Reinitzer went on with the experiment and found another transition at 178.5 °C that absorbed more heat, just like a second melting point. Only at that higher temperature did the liquid finally turn clear. The appearance of what seemed to be two melting points made no sense at all at first. Critics claimed that one of the melting points must be due to an impurity, but Reinitzer protested that his purification was as perfect as it could be. Inspired by the wisdom of interdisciplinary approaches to mysteries that seem to violate the logic of a single discipline, he sought out the physicist Friedrich Ostwald as a collaborator, convinced that there was an explanation to be found within the behaviour of pure cholesterol.

Together they developed the idea that the transition from complete order to complete disorder—from crystalline solid to random fluid—that we term 'melting', or its reverse—the process of 'crystallization', could in principle occur in stages. When a normal fluid crystallizes, each molecule gives up all its ways of moving—its 'degrees of freedom'—at once. In the fluid, any molecule can *be* anywhere and *point* in any direction. But in a crystal it can only be situated at one of the crystal's lattice sites, and it must be oriented in the same sense as all identical molecules at equivalent lattice sites. But what if the ordering of the *positions* and the ordering of *orientations* could be separated into two distinct processes rather than one? And if so, could those orderings be broken down into separate stages that would happen at distinct temperatures?

We tend to put this now in the language of symmetry. A liquid can be moved in space in any of three directions (left-right, front-back, up-down) or a combination of them, and by any displacement, resulting in a state indistinguishable from the original one. When it crystallizes, this symmetry only exists for special displacements that carry the atoms of the crystal a whole number multiple of the crystal lattice spacing. We would say now that the three continuous 'translational symmetries' of the liquid are 'broken' in the solid. There are likewise three angles of continuous rotational symmetry in the normal fluids (think about turning an arbitrary axis through the body of an object though longitude and latitude angles, and spinning it about that axis). In normal crystallization, all these six degrees of freedom, three of translation and three of orientation, are surrendered at once. Their symmetries are all broken at the same temperature. Similarly, they are all liberated at once, at that same temperature, when the crystalline solid melts.

The material of rod-like molecules illustrated in Figure 30 does something radically different. It moves from complete order (the crystal in Figure 30(a)) to complete disorder (the normal fluid—Figure 30(c)) by first freeing the three spatial degrees of freedom (and the one that allows the molecules to spin around their long axis) but keeping two of the rotational angles of freedom restricted. In this intermediate 'semi-melted' phase of Figure 30(b), all the molecules point with their long axis in the same direction (longitude and latitude) just as they did in the crystal, but they are now free to move spatially to any place in the fluid—there is no spatial crystal lattice at all. At the lower melting temperature of cholesterol only the spatial lattice melted away, reacquiring its continuous symmetry but leaving the orientation of the rod-like molecules ordered. Only at the higher transition temperature do the remaining two degrees of rotational freedoms completely 'melt'. Reinitzer and Ostwald called the state of Figure 30(b) a 'liquid crystal' because, like a normal liquid, its molecules are free to move to any position—and the stuff really flows in consequence—but, like a crystal, it can still be further melted into complete disorder.

The chemist and physicist working together really had discovered a new 'phase' of matter, in addition to the three classes of solid, liquid, and gas known from antiquity until that point. They needed a name for the new phase—'nematic' suggested itself from the worm-like nature of the molecular shapes that underlies its emergence. It would be a century before liquid crystals could be recognized and adopted into the larger family of 'soft matter', but the new nematic state displays all the characteristics of soft matter that we listed in Chapter 1. The liquid crystal molecules are built of many atoms; they are mesoscopic objects. Their essential structure lies therefore in their overall rod-like shape, not in their specific atomistic constituents. In consequence, nematic behaviour is universal—there are many examples of molecules which form a nematic state. There is a coarse-grained explanation of their physics, cast in a strikingly beautiful mathematical form, as we shall see. Our theory of liquid crystals takes, as its starting point, a measure of the local orientation of the nematic as a fundamental property, not the individual positions of all the atoms in the fluid. Finally, there are emergent processes and cooperative phenomena with slow dynamics—it is possible through external forces and flows, or just through thermal agitation, for example, to change the direction in which all the molecules point. But this requires the concerted action of many of them, not the independent twisting around of individuals. Such cooperative motion is hugely slower than the motion of single molecules.

A theoretical gift from superconductivity

One of the glories of physics is that ideas and structures from one field frequently open up progress, or provide conceptual tools, in quite another. This is, after all, the reason for the coherence of 'soft matter' in the first place. But such borrowings and mutual illumination have also passed between more distant topics. The 'inverse-square' force/distance laws of both gravity and electrostatics constitute one example—both the gravitational force between two masses and the electrical force from two charges fall off as the square of the distance between them. The common mathematical structure, spotted by Edwards underlying the statistical physics of polymers and the quantum mechanics of particles, is another. Once a problem is solved in one case it is sometimes transposable, with relatively little extra work, to the other. This happy circumstance has occurred at least twice in the story of liquid crystals.

On the first occasion, the Swedish theoretical physicist Lars Onsager became attracted by the similarity between the thermal physics of the rod-like molecules in liquid crystals, and the problem of the orientation of magnetic atoms in a lattice. He had worked on a simplified model for the magnetic system, which shares the properties of a local interaction which tends to align neighbours (magnets in the one case, molecular rods in the other) and thermal motion that tends to destroy the alignment of order (the simple model, due to Ernst Ising, restricts the variation in direction to just two—'up' and 'down', and has enjoyed fruitful application across many fields of physics). In 1949 Onsager was able to demonstrate that, providing colloidal-sized particles were sufficiently elongated, the emergent physics would also show a similarity to that of magnetism in a new phase transition. The disordered liquid (akin to the non-magnetized state of iron where magnetic atoms are unaligned with each other) would give way to an ordered state (mapping on to the magnetized state of iron where all the atomic magnets are aligned) when the rods were sufficiently densely packed. Apart from showing theoretically that an entirely novel phase transition would appear in these liquids, the remarkable aspect to Onsager's demonstration was that this could occur through entirely *repulsive* forces between the molecules—including those that would simply prevent them from overlapping.

In a second significant case of borrowing, ideas from superconductivity and particle physics theory (the so-called 'Higgs mechanism' that drove the search for a new particle at the international Large Hadron Collider in Geneva, and on its discovery the award of the 2013 Nobel Prize in physics to Peter Higgs and François Englert) surprisingly provided a conceptual framework for the theoretical physics of liquid crystals. This 'three-way conceptual borrowing' arises because all these systems exhibit phase transitions, akin to the solid-liquid or liquid-gas transitions, or the magnetic transition of iron, as a function of temperature. In superconductors of the usual kind, below a critical temperature electrons can pair up and fall into an ordered collective state that carries electric current without resistance.

Like all phase transitions, the superconductive transition also changes the symmetry of the material at its bulk level. At high temperatures, the thermal fluctuations at the molecular level are large enough to excite the electrons out of this special cooperative state and into the more disordered 'normal' state. Similarly, at high temperatures in a liquid crystal, thermal motion maintains long-range randomness in orientational direction of the molecular rods. At the macroscopic level the material is a normal fluid—with no preferred orientation over long distances, even though there can be very local orientational order ('short-range order') in the immediate vicinity of a single molecule, since immediate near-neighbours tend to line up. However, as the temperature is lowered, the thermal fluctuations become weaker, and the local interactions between rods that tend to align them are able to drive a transition of the entire system. In the resultant nematic phase of the liquid crystal, a preferred orientation appears at the bulk level. Although there is still no spatial order, so that the system is still a fluid in terms of its propensity to flow, it possesses strong and long-range orientational order.

The imaginative identification of the similarity between these transitions, in liquid crystals on the one hand and superconductors on the other, is due to the French physicist and Nobel Laureate Pierre-Gilles de Gennes. He had been working in the early 1960s at the new national research laboratory at Orsay near Paris, on the type of superconductors that are able to trap magnetic fields inside their ordered phase. His chosen approach involved the clarifying of elegant theoretical tools that had been developed in the 1930s by Russian physicists Lev Landau and Vitaly Ginzburg. Readers interested in its mathematical details may wish to consult the References and further reading section, but the three key ideas behind the 'Landau–Ginzburg' theory of phase transitions can be understood qualitatively and pictorially:

- The first task is to identify a quantity, let us call it S, a property of each point in space that describes how well-ordered the material is in its vicinity. In liquid crystals, this 'order parameter' describes the perfection to which the molecules are locally pointing in the same direction. In superconductors it measures the amplitude of the special conducting quantum state. In liquid crystals $S = 0$ in the disordered state and $S = 1$ in a perfectly ordered nematic state (negative values of S would refer to a different type of 'pancake' ordering of the rods)

- The next step is to identify a 'landscape' of energy in which this parameter changes. This is analogous to a real landscape of hills and valleys—the system will tend to move from high places (of high energy) to low places (where 'place' refers to a value of S) under the jostling of a thermal environment. An example of a set of such (one-dimensional) energy landscapes over a range of values of S is shown in Figure 31.

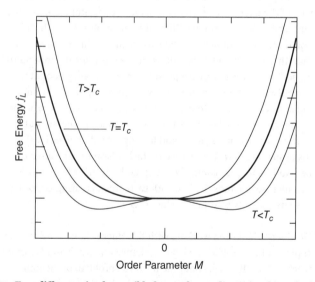

31. Four different, simple, possible forms of a one-dimensional 'Landau' free energy landscape in terms of the degree of order (S) in a liquid crystal undergoing a continuous phase transition. The shape of the landscape is shown at four different temperatures. The local maximum free energy at $S = 0$ appears only for temperatures $T < Tc$, otherwise for $T > Tc$ there is a local minimum at $S = 0$. The bold line indicates $T = Tc$.

- Finally, the landscape itself is allowed to shift, like the geography of a mountain range over aeons of time, but in this case as the *temperature* changes. Valleys at one temperature can become hills at another and vice versa. If the order-parameter value corresponding to a valley bottom shifts, then so will the emergent value of the order itself, as it follows the valleys wherever they appear.

When the liquid crystal system is described using this scheme, in terms of the emergent 'field' of orientational order S, a mathematical similarity to phase transitions in superconducting materials becomes apparent. The similarity can be visualized— Figure 31 presents a simple example of the changing 'energy landscape' (strictly in terms of 'free energy', which measures the part of the energy of a system available to do useful work) for the liquid crystal order at four different temperatures (actually for liquid crystals confined to a flat surface in two dimensions rather than three). At high temperatures, above the 'critical temperature' T_c (so $T > T_c$), the landscape possesses a single valley floor at the value $S = 0$. At temperatures below T_c, however, the minimum in the landscape splits into two, each moving away from zero as the temperature is further decreased (in a symmetric way so that S may be positive or negative) and leaving a small 'hill' or local maximum at $S = 0$. The change in the landscape that causes the new valleys to arise comes from a steadily stronger tendency for the microscopic rods to align with each other, creating attractive low energy 'valleys' at finite values of S.

De Gennes had been working with this 'energy landscape' tool for superconductors when he heard a chance remark from his former teacher, Jean Brossel, about the unsolved problem of the milky appearance of the intermediate phase in liquid crystals. He immediately started to wonder if the general mathematical formulation of continuous phase transitions that he had been working on in a totally different context might have something to say about this old unsolved problem. Since the scattering of light that gives rise to this optical effect can only come from differences in optical properties across distances similar to the wavelength of light, the idea had been advanced that the rod-like molecules were clumping together in some places and diluting themselves in others. The superconductor analogy suggested to de Gennes that he introduce the possibility of slow changes in the principal molecular direction though the liquid, carrying their own elastic-like energy. A much more natural explanation then dropped out—the slow changes in the nematic direction would be enough to scatter light through its changing degree of alignment with the light polarization. The 'clumps' or spatial 'domains' that seemed to have formed were made of particle *orientation*, not of their density. For the second time, a theoretical structure from quantum physics had found itself illuminating a soft matter system.

Emergence of defects—objects made of 'nothing'

The newly borrowed theory's explanation of the milky appearance of the nematic phase was a remarkable insight, but the idea of an 'energy landscape of order' held more surprises. Optical microscopy of liquid crystals of increasing sophistication had revealed the appearance of spatial structures termed 'defects' as well as the simple light scattering from changing order.
Figure 32(a) shows a microscope image of a nematic liquid crystal observed under polarized light in the clever way we anticipated might be revealing in the introduction to this chapter. Light is sent into the sample, with its polarization in the 'left-right' sense, and the emerging light filtered so that only polarization in the 'up-down' direction emerges. If the polarized light is not affected in any way on passing through the sample then no light emerges at all. This is exactly how polaroid sunglasses reduce the glare from reflected sunlight (light becomes polarized through glancing reflection). It is easy to confirm this by taking two polaroid filters and rotating one against the other so that their polarization directions lie at right angles—their appearance turns dramatically

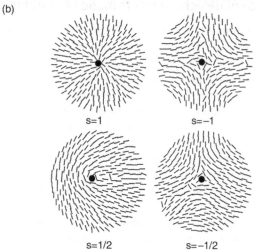

32. (a) Polarized light microscopy revealing a spatial structure of nematic liquid crystal defects of four different 'orders' (field of view is 1 μm) at the meeting point of domains in which the director field points in different directions; (b) the diagrammed patterns of the 'director' field around the different orders of local defect.

black as the pair now allow neither polarization to pass, and give an opaque screen (the sudden black opacity when their polarization directions become perpendicular is always remarkable, regardless of how often it is seen). However, if the material between the two filters possesses a special direction, such as a nematic phase, and if this special direction is neither directly parallel nor at right angles to the polarization of the incoming light, then the polarization of the light can be altered along its passage between the two filters, so that some light does pass through the second. The dark areas of Figure 32(a) indicate that the director field either points directly along or is exactly transverse to the polaroid direction of the first filter—so there is no twisting of the polarization of light passing through the liquid crystal. In other places the orientation of the molecules does generate some twisting of polarization, thus allowing some light to pass.

The optical technique of 'crossed polaroids' reveals a rich and remarkable structure. Spatial changes in orientation are there, to be sure, but they are patterned by point-like structures—the 'defects'—around which the nematic direction changes discontinuously. Careful examination of the patterns of light and dark around these points shows that they possess different 'orders'—there are some with two and some with four bright vanes, for example, in Figure 32(b).

The explanation is contained within the theoretical picture we have already considered. When the temperature falls below the critical value, the orientation of the liquid crystal molecules must adopt a chosen direction—the 'director'—at each point. But that choice is itself symmetric; it is not possible in general to predict which way the orientation will 'fall' from its new local hilltop into the valleys of Figure 31. This naturally gives rise to spatial structures in which the direction of the liquid crystalline order is different at different spatial locations—local domains of orientation, akin to those of a magnet whose magnetization differs at different points. To accommodate such spatial variation of local orientations, very localized 'topological defects' must arise—points at which regions of different orientation fields abut, and where the director cannot be defined. Figure 32(b) illustrates four possible such topological defects in terms of the local behaviour of the orientation field around them. The number, s, describing the order of the defect in each case, refers to the number of rotations that the nematic field undergoes on a close orbit around the defect. Each is identified wherever it occurs in the experimental micrograph in Figure 32(a).

The defect-prone state of Figure 32(a) is in energy minima locally, but it has certainly not found its lowest possible energy *globally*, since this belongs uniquely to the state in which the nematic field is completely uniform, with the same director everywhere. So there must, in principle, be a dynamical route from the defected state to the uniform one. The existence of signs on the defect orders indicates how this might happen—if two defects of opposite sign come into close proximity they can annihilate, leaving no defect at all (just like a particle and its antiparticle in high energy physics). The slow mutual attraction, motion, and annihilation of defects is another delicate example of slow dynamics in soft matter, invoking emergent large-scale objects. In this case, the 'objects' whose dynamics create and define the process are the defects themselves—which are strange in that they are not 'made' of anything at all. Rather they emerge from local stable behaviour of the underlying field, a powerful idea that is shared by particle and superconductivity physics.

The stunning optical structure of Figure 32(a) is only apparent under magnification and using polarized light as well. But the defect structures in liquid crystals scatter ordinary light of mixed polarization indiscriminately—and it is these emergent defects that are responsible for the pearly milkiness of the fluid to the naked eye. Some liquid crystalline materials seem to scintillate with colours too—a clue that other structures are possible.

A zoo of liquid crystal phases

The nematic liquid crystal, in which the only order parameter is the average local molecular rod orientation, turns out to be just the simplest example of a very complex family of materials. Recall that the insight into understanding the nematic phase in the first place was that, in normal crystallization, the set of all three translational and three orientational symmetries in a molecular fluid are 'broken' at once. The nematic phase arose when just two orientational angles (the points on a sphere towards which the nematic director points) were separated from the rest into a distinct group, so that the full transition from liquid to crystal happened in two stages. It is perfectly possible that in some systems, the remaining degrees of freedom separate out further in the distinct temperatures at which they order. For example, there is no law that insists that all three directions of spatial, crystal order should appear at once. It is possible to imagine that a fluid 'crystallizes' along just one direction, creating a series of planes in regular separation, within each of which the system remains fluid in the other two directions.

This is precisely the pattern of another class of liquid crystals known as 'smectic'. The term originates from the Greek word for 'cleansing', as smectic materials have a soap-like appearance. From the insight we already have from the bilayer forming surfactants of real soap, we might guess that smectic materials are also layered. They do indeed introduce one dimension of periodic spatial order in the direction perpendicular to the layers. This new spatial pattern is in addition to the orientational order (so breaking one spatial as well as orientational symmetry). The oriented molecules now lie stacked in the repeated layers, as in Figure 33, which also shows that the new phase can itself assume one of two different types of symmetry. The smectic A-phase displays its molecular rods' director pointed perpendicularly into the layers, while in smectic C-phases the director is tilted at an angle to the layers.

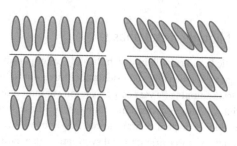

Smectic A Smectic C

33. Schematic illustration of the smectic A- and C-phases.

As the complexity of the phases in liquid crystals increases from nematic to smectic so the number of possible distinct transitions, together with the associated 'melting' temperatures, also increases. A full crystal may first melt into a smectic, then the more disordered nematic, before finally achieving the fully disordered liquid state.

Like nematics, smectic phases also display a defect structure, but in this case a really breath-taking example of naturally occurring geometric complexity. In the same way that regions of different directions of nematic order can be reconciled at their boundaries through 'topological defects' (those of Figure 32), so multiple smectic domains can also be reconciled—but now in structures called 'focal conics'. The cores of the defects around which the smectic layers are curved into partial shells, occur in pairs—one in the closed extended loop of an ellipse, the other intersecting at right angles in the open curve called a hyperbola. Figure 34 provides a visualization of how this happens and a striking photograph of an experimental example.

Another example of the occurrence of such 'conic section' curves in nature is found in the orbital paths of gravitating objects. A captured satellite of a planet or star will in general execute an elliptical orbit, while an object with too high a velocity to be

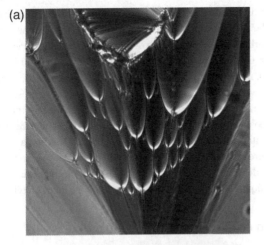

(a)

(b)

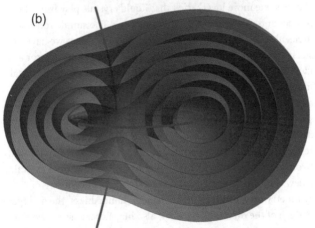

34. (a) Hyperbolic focal conic defects in a smectic liquid; (b) diagrammatic construction of the intersection of elliptical and hyperbolic curves in defects of layered materials.

captured permanently (examples are found in comets that visit the solar neighbourhood from deep space) execute the open orbits of hyperbolae. In a remarkable demonstration of geometric insight, George Friedel was first to identify a smectic liquid crystal in 1910, not by the microscopic observation of the smectic layers (which were too closely spaced for detection by any microscope of the time) but by observation of the conic section forms—the hyperbolae and ellipses—of their defect cores. He knew from prior acquaintance with this advanced geometry that this was a signature of structures built from equally spaced layers and was able to deduce the layer spacing, and so the molecular dimensions of the liquid crystals, without needing to image them directly.

Liquid crystals through the looking glass

There is one more lovely trick that liquid crystals play from the physics and chemistry of their symmetries. The simple rod-like molecules that form the nematic and smectic phases we have met so far possess 'reflection symmetry'—their images in a mirror are identical to the originals. But it requires only very little chemical modification to produce versions that instead share the mirror asymmetry of our left and right hands. Think, for example, of a simple rod with two extensions from its side that emerge from two different places and in different directions—in general, its reflection will not be superposable onto the original. 'Cholesteric' (from Greek again, this time signifying 'coloured') liquid crystal is a new class whose molecules possess this breaking of mirror symmetry. As a result of the molecular 'handedness', the emergent pattern of the director field must also in general possess the same symmetry of 'leftness' or 'rightness.' It achieves this by embodying twist, just like a screw-thread, as in Figure 35.

Note that if you were to travel along the helical axis in Figure 35 you would always encounter an anti-clockwise rotation of the director. Were the molecules to adopt the structure of their mirror images, the twist would be clockwise. The gentle twist creates a

helical axis

pitch *p*

35. Schematic of a cholesteric liquid crystal showing the helical pitch of twist, perpendicular to the director plane.

further aspect to the coarse-grained structure of cholesterics—the newly emergent length scale, p, of the pitch (see Figure 35(b)). This can be much larger than the molecular length itself and commonly falls into the range of dimensions comparable to the wavelengths of visible light (300–700 nm). Here is the physical origin, at second hand, of the name of this 'colourful', rather than 'soapy' (smectic) or 'wormy' (nematic), phase. For white light passing through a cholesteric liquid crystal has those of its wavelengths corresponding to the pitch strongly reflected, through interference of the many weak reflections from each twisted layer. Much like the colours created in the same way from thin oily films on water, or those of peacock feathers and other colourful animals who achieve iridescent colouration by interference of light, rather than absorption, the rich chromaticity from cholesterics is remarkable. The type is named from the very molecule, cholesterol, that Reinitzer was investigating in his first discovery of liquid crystals.

The defect structures of these higher order liquid crystals are correspondingly richer. They also possess similarity mappings with completely different physical systems. So, for example, an analogue of a vortex phase in a (Type II) superconductor that supports a lattice of magnetic field lines was predicted in smectic cholesteric liquid crystals and later found experimentally as their 'twisted grain boundary' phase. The richness comes about from the competing molecular tendencies of layer formation (smectic) and twistedness (cholesteric)—there is no continuous structure that can accommodate both of these. However, by assembling 'grains' of smectic phase and stacking those grains into twisted layers, an accommodation of the two is balanced. De Gennes showed that the mathematics of the defects in this liquid crystal structure mapped beautifully onto that of magnetic lines of flux in the superconductor, when written in the form of fields.

Straight up with a twist

The twisted structure shown in Figure 35, as well as providing a beautiful example of self-assembled structure within ordered soft matter, is also the favoured technological route to the fast, switchable, liquid crystal displays of flat-screen technology. The insight required to understand the simple and effective idea is once more the polarization of light.

Recall that setting two polarizing filters on top of each other, but with the second one, known as the analyser, rotated so that its polarizing direction is perpendicular to the first creates a system that will let no light through at all.

This configuration is that of the 'dark' state of a pixel in a liquid crystal display. The passage of light in the 'bright' state is achieved by a thin nematic layer between the polarizers, designed so that the nematic molecules adjacent to the polarizers' surfaces tend to lie along their polarization direction. The continuity of the director field then induces a gentle twist to the director, connecting the nematic direction at the first polarizer with that of the second (Figure 36(a)). Light passing through the polarizer has its direction of polarization gently turned by interacting with the nematic field as it passes through the liquid crystal, so that by the time it arrives at the analyser it has the correct orientation of its electric field to pass through. The trick of the display technology is to coat the polarizers with a transparent conductor, so that they can also act as electrodes for a voltage source that, when turned on, sets up a steady electric field directed from one polarizer to the other. When this is turned on, it overcomes the gentle twist of the nematic resting state and aligns the director with the field, as in Figure 36(b). Now there is no rotation of the polarization of the

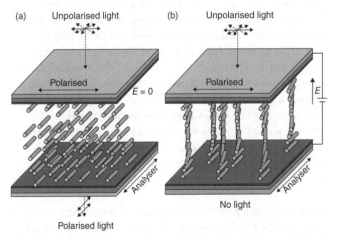

36. (a) A twisted nematic field (nematic molecules shown as cylinders) between two polarizers turns the plane of polarization of transmitted light so that it can pass through both when the electric field is off ($E = 0$); (b) when an electric field turns the director field vertical the light retains the polarization of the first filter and cannot pass through the second.

light passing between the polarizers; it arrives at the analyser with the same polarization with which it left the polarizer and is therefore blocked. The pixel has been switched to dark.

Rods going with the flow

Liquid crystals demonstrate new ways of exploring the idea of 'solidity', so perhaps it is not surprising that, when they are made to flow, their rich structural complexity bursts out afresh into new ways of being a liquid. Identifying how two of the characteristic properties of soft matter systems—special structure at the nanoscopic length scale and the emergence of slow variables—behave in fluid flow will be the key to understanding what happens.

The simplest case of the nematic state generates a strange fluid under the slowest of flows, for as American engineers Jerald Erickson and Frank Leslie showed in elegant theoretical work in the 1960s, there will not be a single 'viscosity' of a fluid liquid crystal, but up to five different numbers that describe how the stress in the liquid responds to the flow rate. The reason for this is the existence of the special direction of the nematic field. While in a simple fluid all shear flows (recall the definition of this special flow in Chapter 1) are equivalent, there are now three quite different ways in which the shear flow direction may be oriented with respect to the nematic field. The nematic director may point along the flow itself, it may point along the direction of increasing flow velocity, or along the 'neutral' direction perpendicular to both.

The rheology of liquid crystals becomes still more interesting as the rate of the flow is increased. The timing clock for flow rate is, as usual, the timescale for internal rearrangement of the soft matter structure—in this case the relaxation time for the orientation of the rods as they diffuse in the Brownian motion of their surroundings. When the flow rate is faster than this reorientation time of the rods (this way of scaling the flow rate is given the name 'Péclet number'), the flow begins to change the orientation of the rods themselves.

The interaction between a shear flow and the nematic director is subtle, and surprising, but at the same time a remarkable mutual achievement of theory and experiment. For both agree on a counter-intuitive series of behaviours as the flow rate is increased. As suggested first by the Italian theoretical soft matter scientist Giovanni Marrucci, the director field first starts to 'tumble'—there is no steady direction, but instead the flow rotates the alignment of the rods (see Figure 37(a)). At higher rates, the field oscillates about a preferred direction (the 'wagging' of Figure 37(b)), before settling down aligned to the flow (Figure 37(c)).

Beautiful experiments by researchers in Leuven, Belgium, in 2005 overcame the materials science challenge of constructing a fluid of identical rod-like molecules at the nanoscale, by purifying a solution of *fd*-viruses that have just this shape and sufficient rigidity. The viruses are nearly 1 micron in length but have a diameter of just 7 nanometres, at room temperature they are effectively rigid, and they are relatively easy to proliferate within the bacteria that they would normally infect. Their static solutions display nematic/cholesteric and smectic phases as the virus particle concentration is increased. It proved possible to use the measurable shear stress of the fluid as a detector of the director motion by first aligning the multiple individual domains of

37. **Illustration of the three dynamic modes of director motion of a liquid crystal (arrowed rod) in a shear flow (shown by flat arrows). Tumbling at low flow rates (a); develops into an oscillatory instability (wagging) at higher rates (b); before flow-aligning (c) at still higher rates.**

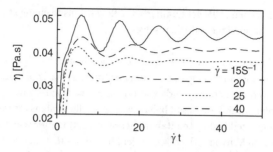

38. Rheological experiments on steady shear flow at four different flow rates of a nematic liquid (in this case an aligned *fd*-virus solution). The y-axis shows the changing shear stress in the fluid, as a function of the total strain (proportional to time) on the x-axis. The oscillations in stress indicate the 'tumbling' behaviour of the director at low flow rates.

differing director angles by a strong flow, then reversing the flow—at this point all the domains would contribute similarly to the stress as their directors would be moving in concert. Measured oscillations of the stress emerge, lessening as the flow rate increases and the director motion migrates though the forms of tumbling, wagging, and aligning (Figure 38).

The Leuven experiments brought more to light—for not only did the shear flow affect the local behaviour of the director with time, but at another larger length scale, the domains and defects were swept along, stretched, distorted, and compressed as well. Theirs is the scale of the wavelength of light, so it affects the visible appearance of the fluid—in other words, pearliness and cloudiness respond to stirring in even more complex ways. The flow experiments display in visibly striking ways the three levels of hierarchical structure that liquid crystals produce—the molecular rods, the director field, and, finally, the pattern of defects and textures, and the subtle ways by which each interacts with the others and the external environment. We have not heard the last from the liquid crystalline realm of soft matter.

Chapter 6
Liveliness

The Japanese research team watched, glued to the microscope eyepiece, as their artificial cell membrane, or 'lipid vesicle', came into focus. The Nagoya researchers had managed to encapsulate quantities of two proteins within the tiny self-assembled bag of surfactant membrane. One of these proteins was known to act as a unit for a self-assembled polymer, so their system constituted an interesting combination of self-assembled soft matter: a one-dimensional polymer contained within a two-dimensional closed membrane. The components jostled about in the field of view under constant Brownian motion. Then something extraordinary happened—the vesicles began to distort from their natural spherical shape, throwing out extensions and becoming highly asymmetric (an example is given in Figure 39). Beyond thermal self-assembly, this new motion had an entirely different character. The soft matter, rather than passively responding to its interactions and the environment in the usual inexorable slide towards thermal equilibrium, was actively moving and changing shape.

The vesicles' shape transformation was reminiscent, if only as a crude shadow, of processes known for many years in living cells. To divide, crawl, and form tissue, biological cells need to change their shape in many ways. The Nagoya research team was one of those attempting to build towards an understanding of these

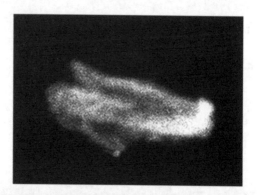

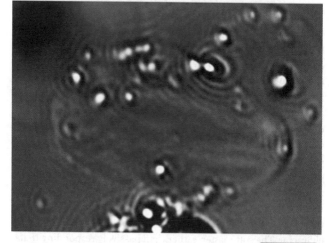

39. **Florescent labelled (top) and white light transmission (bottom) confocal micrographs of vesicles containing polymerized actin and myosin cross-linkers. The scale bar corresponds to 10 microns.**

complex processes by combining just a few of their components within a self-assembled soft matter structure that mimicked a cell membrane—the lipid bilayers of the vesicle wall. They had managed to encapsulate within the vesicle a significant concentration of a protein molecule known as G-actin. Like the peptides of Chapter 4, appropriate solution conditions favour its self-assembly, and by a similar one-dimensional polymerization process. The G-actin units form extended, stiff polymers of filamentous or F-actin.

The key to the research team's special findings was, however, the third ingredient. A cross-linking protein, called myosin, possesses two sub-units that each bind to sites on the F-actin polymer strings—so acting just like the cross-links in a polymer gel (see Figure 40(a)). Myosin is no passive cross-linker, however, rather it draws on energy sources within the cell (and in particular the universal cellular 'fuel' molecule adenosine tri-phosphate (ATP)) to actively pull its two attached filaments past each other. It is an example of a 'molecular motor'. The actin-myosin system forms the basic structure in animal muscle, for example. In muscle, thousands of myosin molecular motors act together in consort to produce a macroscopic muscle contraction. But what the Japanese experiment and many others like it in recent years have shown is that the incorporation of active motor units such as myosin into artificial soft matter assemblies opens up an entire new world of materials, in which the local dynamics are driven chemically rather than simply by thermal motion.

By exploiting the biologically inspired idea of incorporating molecular motors into soft matter systems of polymers, colloids, and membranes, the materials can be propelled far from their equilibrium states. New emergent behaviour results—more recent work has varied the concentration of filaments and motor cross-links in such active gels to find that entirely new structures form, unknown in passive networks. For example, self-organized groups of actin polymers and myosin motors can create ring-like, and star-like ('aster'), formations as the relative concentration of actin increases.

Figure 40(b) and (c) show a fluorescent micrographs of these two structures, formed as the molecular myosin motors 'walk' along

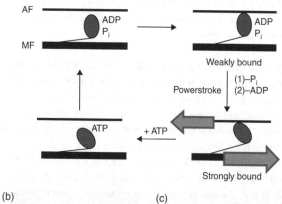

Soft Matter

40. (a) **Schematic of the molecular motor myosin on its own filament (MF) binding to an F-actin filament (AF), and using the chemical transformation of the 'fuel' molecule ATP to ADP to power relative sliding of the two filaments (indicated by block arrows); (b) and (c) two resulting self-organized structures in actin-myosin active polymer gels: (b) rings and (c) 'asters'.**

the actin filaments until the motors cluster at the centre of a radial cluster of polymers.

Almost any type of soft matter we have considered so far possesses an active form. Using myosin (and its molecular 'fuel', ATP) corresponds to making the cross-links of a polymer gel active. But polymerization itself can be actively driven, as well as the cross-linking between polymers. Bacteria are, within this perspective, an active form of colloid—nanoparticles that can swim. As their shape becomes highly anisotropic, they generate the notion of 'active liquid crystals'. In this chapter we briefly review the current excitement of the new biologically inspired field of active soft matter. There is an extra motivation, for as well as exploring further emergent properties of soft matter in 'active' states, they can also be viewed as highly simplified versions of living matter, and so a first step on the long road of understanding the complexity of biology.

Pushy polymers and bacterial propulsion

There are many ways of propelling a vehicle, from engagement with a surface by wheels, legs, or tracks to rowing with oars or rotor pressure against a surrounding fluid, and even rocket propulsion. A less commonly encountered propulsion technique, except perhaps at children's bath time, is the 'soap effect'– grabbing at a slippery bar of soap by one end all too often sends it flying into the air before it is lost in the murky bathwater. A rather subtle combination of the asymmetric, tapered end of the bar, a low friction to sliding, and a lateral squeezing stress from a wet hand, all conspire to create a surprisingly effective driving motion.

More surprising still is that this very effect of 'squeeze propulsion' has been recruited by a bacterium, *Listeria monocytogenes*, as an efficient transport mechanism to serve its infection of host cells. Although text books often give the impression that biological cells are mainly made up of water, surrounded by a lipid bilayer, and containing a few important structures (the nucleus, several proteins, and a sparse cellular skeleton of polymerized biomolecules), the reality is very different. Cells are extremely crowded environments indeed, with water as a minority component. As a consequence, their interiors are very high-viscosity, even viscoelastic, media. Gentle swimming is not going to be an effective propulsion strategy. Instead, *Listeria* seems to have fought elasticity with elasticity—researchers in the 1990s began to notice, using subtle methods in optical microscopy, that behind a moving bacterium there appeared to be a long tail of denser medium (see Figure 41). Analysis revealed that this tail was composed of a cross-linked gel of filamentous F-actin—the same self-assembled polymer used in the Japanese artificial cell mimics. Unlike the propelling oar-like flagella used for swimming by other species of bacteria, such as the laboratory poster-child *Escherichia coli*, the actin tail did not seem to be fixed to the bacterium but instead to the surrounding medium. As the bacterium moves forward, the tail seemed continually to be forming around it. Not only was it using the surrounding host medium as an elastic framework to move through, the invader was taking the surrounding cell's own supply of G-actin to create its propulsion system by building the continuously growing tail from it.

The French soft matter physicist Jacques Prost and his collaborators worked with the biological data to evaluate how the typical *Listeria* swimming velocity of up to 10 microns a minute (see the 5-micron scale bar in Figure 41(a)) could be achieved. Their idea is shown schematically in Figure 41(b). They represented the cell interior as a weak elastic medium, within which a stronger elastic gel is being polymerized at the surface, but only at the rear, of the bacterium. The idea is that by creating new elements of the actin gel at its surface, the bacterium compresses the newly formed section of tail immediately surrounding it against the elastic resistance of the cell interior, so generating compressive elastic stresses there. Because these stresses impinge on the rear of the bacterium only, and not the front, there is a net ('soap-squeezing') force propelling it forward. As it does so, it leaves the actin tail behind in its wake, moving into a yet-unexplored region of the cell where it constructs a new section of tail.

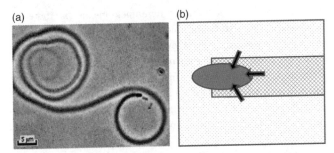

(a)   (b)

41. (a) The path of a *Listeria* moving in a bio-elastic medium, by phase-contrast microscopy, showing the polymerizing actin tail; (b) schematic of the cross-linked actin network (cross-hatched region) and the elastic forces on the bacterium (arrows).

Close inspection of Figure 41(a) shows that the density of monomers in the tail slowly decreases away from the attachment point to the bacterium. This is an important clue to the additional trick played by actin polymerization: the monomers themselves have a 'front' and 'rear' end, and polymerize with a corresponding orientation along their polymers. Furthermore, the F-actin polymers grow at the site that displays bare 'front' ends of the monomers (abutting the bacterium) and tends to fall apart, or 'depolymerize', at the other terminus (far from the bacterium). This is wonderfully useful for the *Listeria* of course—for the continuous dissolving of the extreme rear end of the tail resupplies the cell with raw actin monomer that, after diffusing through the cell, can be reincorporated into the new filament.

Active versus passive self-assembly— polymerization motors

At first glance, the polymerization of the actin filaments and networks that propel the *Listeria* bacterium through its host cell looks very like the self-assembly of peptide fibrils or surfactant worm-like micelles that we encountered in Chapter 4. The monomers possess 'sticky' ends that adhere to each other when thermal diffusion brings them, by chance, in contact, and so the polymers grow. But the thermal environment means that there must be a chance that the polymers break at any point, as well as the reverse process by which they combine. One way of understanding the distribution of polymer lengths that emerges, characterized as we saw by the 'half-length', is that once this mixture is attained, the random breaking, growing, and reforming reactions will not change it further. We say that its length distribution is at 'dynamic thermal equilibrium'.

One of the great results of the science of thermodynamics that grew originally from the steam engineering of the nineteenth century is that a system at equilibrium cannot do work. Yet pushing a bacterium through a high-viscosity cell medium most certainly requires work. For one thing, the friction of the *Listeria* through the fluid generates heat—and that energy requires an energy source. It must be being driven along by a motor of some sort. Although a polymerizing filament of gel does not look much like anything we would normally call a motor, it nevertheless manages to behave as one in the case of *Listeria*. But if an artificial propulsion system of this kind were attempted using the self-assembled peptides, it would not get far. Attaching some short fibrils to the surface of a bacteria-sized particle might move it a short distance while they achieved their equilibrium length, but then there would be as much Brownian pull as push, and the particle would get nowhere. The *Listeria* propulsion seems, in this light, extremely mysterious.

The clue to the puzzle rests in the apparently harmless rule that the actin adds monomers at one end only, while losing them at the other. Although this looks at first sight to be a minor modification of the thermodynamic equilibrium addition and subtraction of monomers from the passive examples of self-assembled polymers, it violates an important consequence of equilibrium known as the 'principle of detailed balance'. Introduced first by the great German pioneer to the microscopic approach to thermodynamics, Ludwig Boltzmann (he of the 'Bolzmann constant' k_B), it was later developed into an important foundational theorem applicable to microscopic states within a system at equilibrium. It is straightforward to define, even if its validity and scope took some of the best minds in science several decades to explore. The important underlying idea is that a large-scale (where the 'mesoscopic' structures of soft matter are still 'large' in this sense) system continually visits examples of the many possible microscopic states to which it corresponds. So the large-scale state of a gas of fixed volume, temperature, and pressure corresponds to many different states of the positions and velocities of its molecules. The principle of detailed balance says that, at thermodynamic equilibrium, the average rate at which the system transitions between one such microscopic state and any other is the same as the reverse rate.

Applying this deep result to the polymerization of actin shows that this cannot be an equilibrium process. The states in this case are simply the positions of actin monomers within the F-actin chain. Adding another monomer onto the growing tip constitutes a jump from one state of assembly to another with one more monomer on the filament. At equilibrium, the rate at which this happens ought to equal the rate for its exact reverse—a step of depolymerization at the tip. But, as we remarked, this is not a property of actin, whose depolymerization happens only at the trailing end of the polymer.

Of course this does not imply that actin polymerization represents an example of a perpetual motion machine. Energy is conserved, and the work that the polymer does in pushing the bacterium along must come from the processing of some form of fuel. Instead, the insight from noticing that the polymerization and depolymerization of actin violates its detailed balance performs two tasks for us: it indicates how the motor properties of self-assembly arise but also points to where to look for chemical changes that drive this process away from equilibrium. The detailed biochemistry of formation of the actin networks turns out to be rather complex, including roles for several other proteins (important examples are the signalling protein WASP and cross-linker Arp2/3), but sure enough, the process calls on the ubiquitous biochemical 'fuel' of ATP, which delivers chemical energy on binding to active molecular machinery of many kinds in the cell, by losing a phosphate group to become adenosine di-phosphate (ADP) (Figure 40(a) indicates, as an example, alongside the shape changes of the motor protein myosin, these chemical stages of the ATP molecular fuel).

Actin polymerization really does constitute an active motor: it consumes biochemical fuel and can exert a continuous force within networks, and it can even propel cargo, like the *Listeria* bacterium. Like all motors, it has a finite capacity and, given sufficient resistance, will grind to a halt. More recent experiments have shown that, by setting up the right biochemistry, actin tails can be made to propel synthetic beads as well as bacteria. Working with such artificial cargo opens up investigations of the strength and power of the actin motor, for laser optics techniques can now be used to manipulate the nano-beads individually with microscopic resolution. Laser-light 'optical traps' can deliver forces directly onto a bead being propelled by a polymerizing actin tail. When this force is directed against the propulsion, the growth can indeed be slowed down and stopped.

Actin polymerization moves cells in other ways than through the external *Listeria* tails. Cells that require mobility on surfaces have developed a way to extend their leading edge (and also to adhere it to the substrate) by internal active actin polymerization. Figure 42 shows a beautiful example of this.

Active colloids

Since the properties of active polymers are so radically different from their equilibrium counterparts, we might expect to discover as distinct a difference in the case of active forms of colloidal soft matter. There are, as in the polymer case, important direct

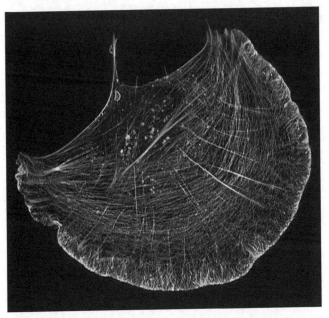

42. A fibroblast cell (one that creates extra-cellular tissue, e.g. in wound-healing) crawling on a surface (towards bottom right) with the internal actin network visualised by fluorescence microscopy.

biological examples of this notion in collections of swimming bacteria, but there are also much simpler, synthetic systems that demonstrate completely novel behaviour.

A delightful example of a form of motor-driven colloid has emerged recently from the work of several laboratories. Colloidal silica or polystyrene spheres are coated on one side with a material that catalyses the decomposition of hydrogen peroxide (H_2O_2; a common bleaching agent) dissolved in water. The well-known bleaching reaction produces oxygen, and when this occurs preferentially on the coated side of the particles, it delivers a net thrust to the particle through the difference in concentrations of peroxide set up by the reaction between their fronts and backs.

The driving force persists while the supply of peroxide lasts. One might conjecture that, given the constant direction of the force on them, the particles must travel in perfectly straight lines, but this is not the case. Brownian random forces are still at work, including those that randomly rotate the particles. The direction of their scuttling motion is therefore always changing, and at first sight does not appear very different from thermal Brownian motion.

An additional functionality has recently been added to this experimental system by Paul Chaikin's group in New York. The catalytic site on one side of the spherical polymer particles is now a microscopic cube of hematite crystal (see Figure 43(a)). The advantage of this material is that its catalytic action to break up the hydrogen peroxide only applies under illumination by blue light and on particles floating at the surface of their container. The experimenter can therefore turn the driving force on and off at will, observing a two-dimensional system of controllable active colloids. Under the active propulsion, the colloidal particles begin

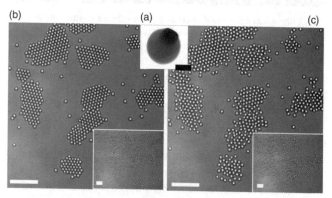

43. (a) Scanning electron microscopy (SEM) of the bimaterial colloid: a polymer colloidal sphere with protruding haematite cube (dark); (b) living crystals assembled from a homogeneous distribution (inset) under illumination by blue light; (c) living crystals melt by thermal diffusion when light is extinguished: Image shows system 10 s after blue light is turned off. The scale bars indicate 10µm in every case.

to collide. After a while larger persistent clusters with crystalline order emerge, themselves colliding and combining as time goes on (Figure 43(b)). That there is no equilibrium attractive force between the particles can be checked by turning the driving force—the blue light—off, after which the clusters simply break up and disperse under thermal motion.

It seems that driven systems can display an apparent attractive force when none exists at equilibrium, purely as a consequence of their dynamics. The phenomenon exhibited in clear detail by the New York experiments can be understood in a simple way by thinking about the mobilities of particles when free and in a cluster. When surrounded by a high density of other particles, the trajectories of any given particle are highly constrained; its effective spatial diffusion is much lower than when it is free. Such regions of low mobility attract clusters of moving agents in general, simply because they all spend more time in them, on average, than in low-density regions where they move faster. In just the same way that shoppers in a high street slow down to gaze at an attractive window display, so forming a local cluster, so the particles slow each other down as they come into contact. When they are all spheres of the same size, the tendency to pack as densely as possible favours the creation of an ordered crystal.

There are some intriguing technological and medical applications for controllable active colloids. As therapeutic delivery agents to exact locations, they seem potentially very attractive. The challenge lies not only in the development of more benign propulsion systems than the highly oxidizing H_2O_2 solutions but also in making the colloids steerable.

Bugs in the service of science: bacteria as active colloids

A lesson might be learned here from nature's own 'active colloids'. We have already met with *Listeria monocytogenes* and its external polymeric motor of driven actin assembly. Other species have developed propulsion systems that they carry with them, alongside a number of strategies for steering. The commonly studied gut bacterium *E. coli* possesses a number of spirally structured flagella that are attached to its outer membrane by protein clusters that themselves constitute rotary motors. These miracles of molecular evolution are a fascinating topic in themselves, driven by proton chemical gradients themselves set up by the processing of the biological 'universal fuel' of ATP we encountered as the power source of the actin gel motor in *Listeria*. For our purpose, we need only know that, in response to signal pathways interior to the bacterium, the rotors and their attached flagella may rotate in either direction. Spinning clockwise forces the flagella to splay out from the bacterial body in all directions, but reversing the direction of rotation to anticlockwise mutually engages their helical forms—they bundle behind the bacterium cell body, forming a rotating corkscrew that propels *E. coli* forward. This bundling is itself another interesting example of an effective attractive force between driven elements out of equilibrium (see Figure 44).

The two states of rotation elicit completely different types of motion in the bacterium itself. When the flagella bundle, they result in a

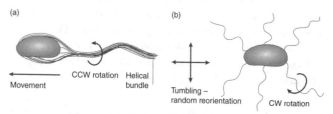

44. The flagellar propulsion system of the bacterium *E. coli*. The direction of flagellar rotation determines straight swimming or tumbling motion: (a) anticlockwise rotation bundles the helical flagellae and propels the bacterium; (b) clockwise rotation disaggregates the flagellae and causes the bacterium to tumble.

period of driven, roughly straight motion along the axis of the cell. This corresponds to the periods of the directed running motion of an active colloid in a catalysed peroxide solution before Brownian motion changes its direction appreciably. When randomly splayed under clockwise rotation, however, the average force on the bacterium is zero; instead it rotates randomly. This is a driven form of the random Brownian rotation on the colloids (which is also, of course, ever-present in the case of bacterial cell as well).

Surprisingly, control over these two dynamical states alone permit *E. coli* to swim deterministically towards sources of nutrients and away from toxins (the process is called 'chemotaxis'). The essential subtlety is a variation in the rate at which it switches from 'run' to 'tumble' modes. The biochemical receptors and their internal biochemistry are complex but have been elucidated over a fifty-year research programme. The emergent rule is astonishingly simple: if the bacterium has recently moved in a favourable direction, then it should tumble less frequently; if the environment is becoming less welcoming, then it should tumble more. Although each tumbling phase randomizes the subsequent direction of run completely, the linking of the randomization *rate* to the recent experience of gradients in nice or nasty ingredients produces a net drift of *E. coli* in the favourable direction. Again there is a lesson in the difference between active and equilibrium systems: just one active component in a chain of dynamic processes is able to alter behaviour completely.

A similar chemotactic drift has been observed in a version of the hydrogen peroxide artificial system that uses rod-like particles with different metal coatings on each end. In this case, the speed of driven motion depends on the local H_2O_2 concentration. So although the Brownian 'tumbling' process does not change its rate in this case, the ratio of run speed to tumble rate does depend on a local chemical gradient—the particles swim towards higher concentrations of the active ingredient. This may suggest other, simply applicable technologies for steering active colloids.

Active rheology and spontaneous flow

We saw in the case of ordinary passive colloids that, as well as the individual particle dynamics of thermal diffusion, fluids containing many particles possessed properties beyond those of the suspending fluid. The viscosity rises with the concentration of particles, for example, and in the limit of high concentration may form gels or even crystals. Slow dynamics emerge from the large-scale configurations of colloidal particles that may even impart weak viscoelastic stresses to the combined fluid. Strong shear flows create larger string-like aggregates that can even arrest the flow completely, as in the corn starch effect. The same questions of emergent properties can be asked when the particles are active. An entire sub-field of rheology has coalesced in the last few years (at the time of writing): 'active rheology' is intensely active itself, opening up a universe of non-equilibrium, self-driving fluids with their own local energy sources. We have space here to glance at only a few of the surprising and profound ideas that have emerged from the new research programme.

The simplest rheological experiment, as we have noted several times in the classical probing of the softness of soft matter, is the measurement of viscosity. Fluid is placed in a shear cell, and the stress on sliding plates, or nested cylinders, depending on the exact geometry of the experiment, is measured as the rate of the shear flow is increased. The viscosity is defined as the ratio of shear stress to shear rate. Any departures from a constant value indicate that the fluid is changing its structure significantly from its equilibrium state. In the case of bacterial suspensions, very much smaller volumes of fluid only are typically available, and 'microfluidic' techniques are necessary.

'Active fluids' may be generated by suspending finite concentrations of swimming bacteria, such as the corkscrew-propelled *E. coli* of chemotaxis, but it turns out that this favourite

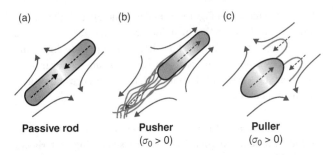

Passive rod   **Pusher**   **Puller**
               $(\sigma_0 > 0)$ $(\sigma_0 > 0)$

45. **Basic mechanism for the active contribution to the rheology in a uniform imposed shear flow (defined in Figure 3, Chapter 1), comparing the case of a pusher and a puller to that of a passive rod-like particle. Flow-induced forces exerted by a passive particle (dashed arrows in (a)) and active forces exerted by the swimmers on the fluid as a result of self-propulsion (dashed arrows in (b) and (c)) drive local disturbances to the flow (solid arrows). In the case of a 'pusher' such as *E. coli*, the disturbance enhances the applied flow, thus reducing the apparent viscosity; the effect is reversed for a 'puller' such as *C. reinhardtii* ($\sigma_0 > 0$).**

lab bacterium does not represent the general case. For the first rheological lesson in active suspensions is that they come in two types, due to the two fundamental ways in which a bacterium can swim. *E. coli* is a 'pusher'—its motor drives it, like a nanoscopic outboard motor, from behind. There are other bacteria whose flagella emerge from their front end and execute swimming strokes that *pull* them through the surrounding fluid. The classic example is *Chlamydomonas reinhardtii* (see Figure 45 for a schematic of its flagella and the flow field they set up when swimming, in contrast to that of an inert rod-like particle and *E. coli*).

At first glance, the fluid flow around pushers and pullers might be thought very similar—after all, both are self-propelled, rod-like particles moving through a viscous medium. However, a downstream centre of propulsion drives an oppositely directed flow of the fluid (relative to the particle's motion) to the case when the centre of propulsion is upstream, as shown in the two schematic diagrams in Figure 45. Pullers generate a similar pattern of extra flow to that of an inert rod, while pushers create a perturbed contribution to the flow field in the opposite direction. Pullers draw fluid in from front and back, pushing it outwards from their sides, while pushers draw in from the sides and expel in front and behind.

The consequences for the measured viscosity in a shear flow containing a significant density of actively swimming bacteria are remarkable. As one might expect, the pullers, with similar extra flow field to that of inactive particles, increase the viscosity as they swim, but to an enhanced degree. The big surprise comes in the case of pushers like *E. coli*: they *reduce* the viscosity, in some cases to only 30 per cent of its background value. The bacteria are therefore acting as a sort of 'active lubricant', aligning with the shear flow and paddling along with it. Less work is required from the external sliding boundaries to generate the shear flow. As the flow rate increases, so the finite additional velocities contributed by the bacterial become less significant, and the viscosity approaches once more that of the background fluid. The observation of very low viscosities raises another intriguing question—might its effective value pass through zero and become negative? In that case flow would need no external force to generate it at all: we would have a fluid that generated spontaneous flow within itself.

Active soft superfluidity and active nematics

Active components, when at high density in a fluid, do indeed give rise to spontaneous flows. Increasing the activity or density, or decreasing the confinement of the channel of an active fluid all predispose it to self-flowing structures of increasing complexity. Figure 46 shows a flow field of a solution containing actively swimming *E. coli* at high density in a small section of a microchannel. Waves of patterned flow have arisen, even breaking the symmetry of the channel. At this value of activity, the flow is not constant but periodic. At higher activities still, the flow becomes chaotic, mimicking the phenomenon of turbulence observed commonly in rising smoke, tumbling streams of water falling into pools, and billowing clouds. The new, active fluid type of turbulence has a very different origin from these everyday examples, however.

In normal turbulence, the transition from steady, smooth flow to chaotic motion is tripped by the increasing dominance of *inertia* relative to viscous forces, as the mean velocity increases. That is, the momentum of the moving and massive fluid carries it into the whorls and eddies of chaotic flow. The key idea is that these inertial forces grow faster than the viscous ones with increasing speed (such an out-pacing effect is usually termed 'non-linearity'— doubling the flow speed more than doubles the inertial forces). In the case of laboratory active fluids of bacterial suspensions, however, inertia is entirely negligible—passive fluids in these conditions never show chaotic behaviour. Yet the presence of activity reintroduces a similar non-linearity into the fluid motion. Using the standard soft matter technique of coarse-graining has suggested models for active fluids that work with scales much larger than the size of individual bacteria, but instead with local volumes containing thousands of them. The relevant variables are the mean swimming direction, the activity, and the fluid velocity itself.

Although this approach loses sight of the individual agents responsible for the fluid properties, by averaging over them, it keeps track of the essential emergent physics of the fluid. The idea is analogous to the derivation of mathematical viscoelastic flow models from the picture of entangled polymer chains, which retain their key properties while losing sight of individual polymers. One of the key coarse-grained variables we have met before—the local orientation and its strength—was an emergent aspect of theories for ordinary liquid crystals. So dense bacterial suspensions, when the bacteria themselves are elongated, furnish

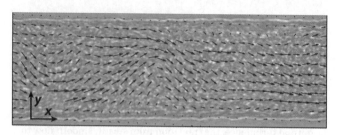

46. Experimental flow field of an *E. coli* suspension in a channel.

the final active version of the soft matter examples we have
introduced—active nematic liquid crystals. Computer simulations
using versions of coarse-grained nematic theories, such as those
we saw in Chapter 5, but enhanced by 'active' terms absent from
the equilibrium case, have produced promising patterns of
self-organized flow, closely patterning experiments such as those
shown in Figure 46, even including transitions to chaotic motion.

Entirely new patterns of behaviour in active soft matter are still
appearing with almost every new experiment. It is likely that
whole classes of phenomena remain to be discovered. Soft matter
science started in the 20th century by providing a long-sought
window onto the reality of the molecular world. It promises, a
century later, to open up new insights into the biological world. It
is time to sum up where this short introduction to its rich history
and current questions has brought us.

Conclusion—from soft matter to life

Two of the great pioneering physicists of soft matter wrote occasionally, but in self-effacing and characteristic modes, about their approach to science. De Gennes quoted (in translation) the French poet Eugène Boudin in his Nobel Prize lecture:

> Have fun on sea and land
> Unhappy it is to become famous
> Riches, honors, false glitters of this world
> All is but soap bubbles

reminding scientists to retain the playfulness of curiosity and perhaps to avoid the poison of self-importance. The shimmering soap bubble and its teasing questions—why a sphere? Why the colours? Why when it bursts only tiny droplets? All science starts with the visible and tangible human encounter with nature, but then leads onwards, into the invisible worlds of the mind, where we now visualize the bilayers, the hydrophobic forces, the self-organization, all revealed by multiple reflections of impinging light or the mysteriously wave-like neutrons. This brief taste of some of the fields of soft matter has followed suit, glimpsing how milkiness draws us to the idea of colloids, sliminess to polymers, pearliness to liquid crystals, and the spontaneous formation of soap bubbles to self-assembly.

Edwards, in a 1973 article in the popular science journal *New Scientist*, advises on the next step, knowing from experience that many of the best ideas are borrowed and adapted rather than forged anew:

> Euclid doesn't really apply to science. The whole is greater than the sum of the parts. If you can manage to study several subjects, you find there are fruitful relations between them, and you find some problems have been solved in some subjects, and you can solve the same problem in a different subject as a theorist—you have only to interpret the symbols in a different way. One can be a very good entrepreneur in this business.

The scientific imagination cannot work in a vacuum, and feeding the mind not only on the problems immediately to hand but on all manner of possible fields of solution, even unlikely ones, is as good advice to a young scientist now as it was then. Edwards had already, at the point of the interview from which the quote comes, experienced the transplantation of quantum field theory into the statistical mechanics of polymers and networks. There must be, as others have commented, something lurking deep in the structure of space and time itself that allows these sort of moves to be made.

This book's skim over soft matter has left far more unsaid than said. But that leaves great riches in store for readers who wish to learn more. Recent developments in rheology at the single molecule level through atomic force microscopy, and the single polymer imaging of confocal microscopy would take us into more local detail. The materials science of fracture, strength, and toughness has hardly received a mention, but it meets soft matter frequently at the science and technology of advanced composite materials, some of them self-assembled. Only 'soft' in a certain sense, the field of granular materials and powders, their flow, the stability and the stress-distribution of their assemblies, has been spawned from soft matter, and it has yielded crops of new science for which we have had no space here. Likewise, the science of food is a rapidly developing branch of soft matter that shares with everything we have met so far the double richness of technological application and scientific bounty.

The final chapter has pointed to the truth that, through evolution, life itself has recruited the structures of soft matter for its own purposes. Polymers, membranes, liquid crystallinity, self-assembly form its fundamental constituents, from the humblest bacterium to the most complex multicellular creature. The tree of life has taken these materials and combined them in vast complexities of such intricacy that the student of simple systems is easily bewildered by the questions of life. Our brief foray into 'active matter' is but a shadow of its realization in living cells. But the tale of soft matter is one of the success of collaborations, of building teams of people with different approaches and skills. There is every likelihood that the next generation of scientists will need to blur the distinctions between 'chemist', 'physicist', 'biologist', and even 'computer scientist' that we have grown up with, and learn to meet the material world on its own terms, rather than on ours.

References and further reading

Chapter 1: The science of softness

Lucretius (1977) *On the Nature of Things*, trans. F.O. Copley. London: Norton, p. 38

P.W. Anderson (1972) 'More is Different,' *Science*, **177**, 393–6

R. Brown (1828) 'A Brief Account of Microscopical Observations Made in the Months of June, July, and August, 1827, on the Particles Contained in the Pollen of Plants; and on the General Existence of Active Molecules in Organic and Inorganic Bodies,' *Edinburgh New Philos. J.* **5**, 358–71; reprinted in *Philosophical Magazine* (1828) **4**, 161–73. This article and the addendum (1829) 'Additional Remarks on Active Molecules,' *Edinburgh Journal of Science* **1**, 314 can be downloaded: https://sciweb.nybg.org/science2/pdfs/dws/Brownian.pdf

A. Einstein (1905) 'On the Movement of Small Particles Suspended in a Stationary Liquid Demanded by the Molecular Kinetic Theory of Heat,' free access: https://www.zbp.univie.ac.at/dokumente/einstein2.pdf [Ger. orig. 'Über die von der molekularkinetischen Theorie der Wärme geforderte Bewegung von in ruhenden Flüssigkeiten suspendierten Teilchen,' *Annalen der Physik* **17**, 549]

Chapter 2: Milkiness, muddiness, and inkiness

Proceedings of the Royal Institution (1839) Accessible in print at the Royal Institution, Albemarle St, London

A. Einstein (1905) 'On the Movement of Small Particles Suspended in a Stationary Liquid Demanded by the Molecular Kinetic Theory of Heat,' free access: https://www.zbp.univie.ac.at/

dokumente/einstein2.pdf [Ger. orig. 'Über die von der molekularkinetischen Theorie der Wärme geforderte Bewegung von in
ruhenden Flüssigkeiten suspendierten Teilchen,' *Annalen der
Physik* **17**, 549]

M. Denn et al. (2018) 'Shear Thickening in Concentrated Suspensions
of Smooth Spheres in Newtonian Suspending Fluids,' *Soft Matter*
14, 170–84

B.M. Guy et al. (2015) 'Towards a Unified Description of the Rheology
of Hard-Particle Suspensions,' *Physical Review Letters* **115**, 088304

M. Wyart and M. Cates (2014) 'Discontinuous Shear Thickening
without Inertia in Dense Non-Brownian Suspensions,' *Physical
Review Letters* **112**, 098302

J.G. Kirkwood (1939) 'Molecular Distribution in Liquids,' *The Journal
of Chemical Physics* **7**, 919

L. Papadopoulos et al. (2018) 'Network Analysis of Particles and
Grains,' *Journal of Complex Networks* **6** (4), 485–565. https://doi.
org/10.1093/comnet/cny005

J. Perrin (1910) 'Brownian Movement and Molecular Reality,'
trans. F. Soddy. London: Taylor and Francis; original paper,
'Le mouvement brownien et la réalité moléculaire' (1909)
appeared in the *Annales de chimie et de physique* **18** (8me
Serie), 5–114

Chapter 3: Sliminess and stickiness

M. Doi and S.F. Edwards (1986) *The Theory of Polymer Dynamics*,
Oxford: Oxford University Press

R.P. Feynmann and A. R. Hibbs (1965) in emended edn: D.F. Styer
(2005), *Quantum Mechanincs and Path Integrals*. New
York: Dover

M. Rubinstein and R.H. Colby (2003) *Polymer Physics*. Oxford:
Oxford University Press

H. Staudinger (1920) 'Über Polymerisation' *Berichte der Deutschen
Chemischen Gesellschaft. A/B* **53**, 1073–85. Free-access: https://
onlinelibrary.wiley.com/doi/10.1002/cber.19200530627

Chapter 4: Gelification and soapiness

A. Aggeli et al. (2001) 'Hierarchical Self-assembly of Chiral Rod-like
Molecules as a Model for Peptide β-sheet Tapes, Ribbons, Fibrils,

and Fibers,' *Proceedings of the National Academy of Sciences* 98, 11857–62

V. Castelletto and I.W. Hamley (2004) 'Morphologies of Block Copolymer Melts,' *Current Opinion in Solid State and Materials* 8 (6), 426–38

M.E. Cates and S.J. Candau (1990) 'Statics and Dynamics of Worm-like Surfactant Micelles,' *Journal of Physics: Condensed Matter* 2, 6869

Ehud M. Landau and Jürg P. Rosenbusch (1996) 'Lipidic Cubic Phases: A Novel Concept for the Crystallization of Membrane Proteins,' *Proceedings of the National Academy of Sciences* 93 (25), 14532–5

Chapter 5: Pearliness

Friedrich Reinitzer (1888) 'Beiträge zur Kenntnis des Cholesterins,' *Monatshefte für Chemie* 9, 421–41

P.G. de Gennes and J. Prost (1993) *The Physics of Liquid Crystals.* Oxford: Clarendon Press

F.M. Leslie (1968) 'Some Constitutive Equations for Liquid Crystals,' *Archive for Rational Mechanics and Analysis* 28, 265–83

M.P. Lettinga et al. (2005) 'Flow Behavior of Colloidal Rodlike Viruses in the Nematic Phase,' *Langmuir* 21 (17), 8048–57.

A.M. Menzel (2015) 'Tuned, Driven and Active Soft Matter,' *Physics Reports* 554, 1–45

L. Onsager (1949) 'The Effects of Shape on the Interaction of Colloidal Particles,' *Annals of the New York Academy of Science* 51, 627–59

Chapter 6: Liveliness

F. Backouche et al. (2006) 'Active Gels: Dynamics of Patterning and Self-organization,' *Physical Biology* 3, 264

S.M. Butler and A. Camilli (2005), 'Going against the Grain: Chemotaxis and Infection in Vibrio Cholerae,' *Nature Reviews Microbiology* 3 (8), 611–20

F. Gerbal et al. (2000) 'An Elastic Analysis of Listeria Monocytogenes Propulsion,' *Biophysical Journal* 79, 2259–75

J. Palacci et al. (2013) 'Living Crystals of Light-activated Colloidal Surfers,' *Science* 339, 936–40

D. Saintillan (2018) 'Rheology of Active Fluids,' *Annual Review of Fluid Mechanics* **50**, 563–92

K. Takiguchi et al. (2008) 'Entrapping Desired Amounts of Actin Filaments and Molecular Motor Proteins in Giant Liposomes,' *Langmuir* **24**, 11323–6

H. Wioland et al. (2013) 'Confinement Stabilizes a Bacterial Suspension into a Sprial Vortex,' *Physical Review Letters* **110**, 268102

Index

For the benefit of digital users, indexed terms that span two pages
(e.g., 52–53) may, on occasion, appear on only one of those pages

NOTHING
A Very Short Introduction
Frank Close

What is 'nothing'? What remains when you take all the matter
away? Can empty space - a void - exist? This *Very Short
Introduction* explores the science and history of the elusive void:
from Aristotle's theories to black holes and quantum particles,
and why the latest discoveries about the vacuum tell us
extraordinary things about the cosmos. Frank Close tells the story
of how scientists have explored the elusive void, and the rich
discoveries that they have made there. He takes the reader on a
lively and accessible history through ancient ideas and cultural
superstitions to the frontiers of current research.

'An accessible and entertaining read for layperson and scientist
alike.'

Physics World

SUPERCONDUCTIVITY
A Very Short Introduction
Stephen J. Blundell

Superconductivity is one of the most exciting areas of research in physics today. Outlining the history of its discovery, and the race to understand its many mysterious and counter-intuitive phenomena, this *Very Short Introduction* explains in accessible terms the theories that have been developed, and how they have influenced other areas of science, including the Higgs boson of particle physics and ideas about the early Universe. It is an engaging and informative account of a fascinating scientific detective story, and an intelligible insight into some deep and beautiful ideas of physics.

CHAOS
A Very Short Introduction
Leonard Smith

Our growing understanding of Chaos Theory is having fascinating applications in the real world - from technology to global warming, politics, human behaviour, and even gambling on the stock market. Leonard Smith shows that we all have an intuitive understanding of chaotic systems. He uses accessible maths and physics (replacing complex equations with simple examples like pendulums, railway lines, and tossing coins) to explain the theory, and points to numerous examples in philosophy and literature (Edgar Allen Poe, Chang-Tzu, Arthur Conan Doyle) that illuminate the problems. The beauty of fractal patterns and their relation to chaos, as well as the history of chaos, and its uses in the real world and implications for the philosophy of science are all discussed in this *Very Short Introduction*.

'... Chaos... will give you the clearest (but not too painful idea) of the maths involved... There's a lot packed into this little book, and for such a technical exploration it's surprisingly readable and enjoyable - I really wanted to keep turning the pages. Smith also has some excellent words of wisdom about common misunderstandings of chaos theory...'

popularscience.co.uk

SCIENTIFIC REVOLUTION
A Very Short Introduction
Lawrence M. Principe

In this *Very Short Introduction* Lawrence M. Principe explores the exciting developments in the sciences of the stars (astronomy, astrology, and cosmology), the sciences of earth (geography, geology, hydraulics, pneumatics), the sciences of matter and motion (alchemy, chemistry, kinematics, physics), the sciences of life (medicine, anatomy, biology, zoology), and much more. The story is told from the perspective of the historical characters themselves, emphasizing their background, context, reasoning, and motivations, and dispelling well-worn myths about the history of science.

www.oup.com/vsi